A

Abamectin [71751-41-2], 24:830
Abbe constant
  of vitreous silica, 21:1057
Abernathyite [12005-93-5], 24:643
Abex
  alkylphenol surfactant, 23:503
Abherents, 21:207
Abhesives, 21:207
Abietic acid [514-10-3], 21:294
Ablative materials
  vitreous silica fibers in, 21:1065
Abrasion index, 22:228
Abrasives
  particle shape classification in, 22:267
  PVB in, 24:935
Abrupt interfaces, 21:775
Abrusoside D [125003-00-1]
  potential sweetener, 23:573
ABS. See Acrylonitrile–butadiene–styrene
    copolymer.
Absorption
  of hydrogen sulfide, 23:435
  role in sulfuric acid prdn, 23:379
  simultaneous heat and mass transfer,
    22:198
Absorption centers
  in vitreous silica, 21:1061
Accelerators
  for butyl rubber, 23:805
  for grouting systems, 22:454
  in latex products, 21:584
  rubber chemicals, 21:460
  sodium nitrite use in, 22:400
Accelerometers
  sensors for, 21:820
Accroides, 21:298
Accugel, 24:1079
Accuspin, 22:80
Acentric factor, 23:1012
Acepromazine [61-00-7], 24:834
Acesulfame [33665-90-6], 23:563
  compared to sucrose, 23:4
Acesulfame-K [55589-62-3]
  aspartame with, 23:559
  as sweetener, 23:563
Acetalization
  of poly(vinyl alcohol), 24:925, 991
Acetals
  from sugar alcohols, 23:102
p-Acetamidophenyl salicylate [118-57-0],
  21:615

Acetaminophen
  from SCF solutions, 23:472
Acetic acid [64-19-7]
  solubility in steam, 22:730
  specific gravity, 23:625
Acetobacter aceti
  inhibited by sorbates, 22:580
Acetobacter xylinum
  inhibited by sorbates, 22:580
Acetochlor, 22:422
Acetone [67-64-1]
  dielectric constants of, 22:759
  as solvent, 22:546
  supercritical fluid cosolvent, 23:457
Acetonitrile [75-05-8]
  as HAP compound, 22:532
  as solvent, 22:546
Acetoquat
  surfactant, 23:529
o-Acetoxybenzoic acid [50-78-2], 21:617
3η-Acetoxy-19-hydroxy-5-cholestene
    [750-59-4], 22:879
Acetoxymethylation
  tellurium dioxide as oxidant, 23:786
Acetoxysilanes, 22:87
Acetylacetone, 24:291
Acetyl chloride [75-36-5], 24:855
Acetylcholine, 24:830
Acetylcholinesterase, 24:508
  inhibition of, 24:830
Acetyl-CoA carboxylase, 23:751
Acetylene [74-86-2], 24:852
  use in optical spectroscopy, 22:645
(17α)-17-(Acetyloxy)-13-ethyl-18,19-
    dinorpregn-4-20-yn-3-one, 22:903
Acetylsalicylic acid [50-78-2], 21:604, 617
4-O-Acetylsucrose [63648-80-6], 23:68
6-O-Acetylsucrose [63648-81-7], 23:66
2-Acetylthiophene [88-15-3], 24:43, 45
Achromobacter sp.
  inhibited by sorbates, 22:580
Acid Blue 3 [3546-49-0], 24:556
Acid Blue 9 [2650-18-2], 24:559
Acid Blue 48 [1324-77-2], 24:556
Acid Blue 83 [6104-59-2], 24:566
Acid Blue 93 [28983-56-4], 24:556
Acid Blue 119 [1324-76-1], 24:556
Acid dyes
  in wood stains, 22:692
Acid Green 5 [5141-20-8], 24:556
Acid hydrolysis
  of starch, 23:599

Acid-modified starches, 22:713
Acid rain
    effect of sulfur dioxide on, 23:253
Acids
    tanks for, 23:644
α-Acids
    from hops, 23:466
Acid Violet 6B [1694-09-3], 24:564
Aclame, 23:570
Acoustic control
    shape-memory alloys for, 21:972
Acoustic measurements
    viscoelastic properties, 21:412
Acoustic waves
    sensors based on, 21:823
Acronal, 24:1064
Acrylamide [79-06-1]
    in grouting systems, 22:454
Acrylamide grouts, 22:454
2-Acrylamido-2-methylpropanesulfonic
        acid [15214-89-8], 23:208
2-Acrylamidopropanesulfonic acid
        [33028-26-1], 23:208
2-Acrylamido-2-pyridylethanesulfonic acid
        [79647-74-8], 23:208
2-Acrylamido-2-(p-tolyl)ethanesulfonic
        acid [79647-73-7], 23:208
3-Acrylamido-2,4,4-trimethylpentane
        sulfonic acid [79647-72-6], 23:208
Acrylate-acetate copolymers
    for hydroseeding, 22:461
Acrylate grouts, 22:455
Acrylates
    from poly(vinyl alcohol), 24:990
Acrylic acid
    polymerization in SCFs, 23:470
    VP copolymerization, 24:1090
Acrylic elastomers, 21:490
Acrylic fibers
    dyes for, 24:551
Acrylic polymers, 21:490
Acrylic resins
    thiols in, 24:28
Acrylics
    in tire cord, 24:164
Acrylonitrile
    copolymerization with VDC, 24:888
Acrylonitrile-butadiene-styrene
        copolymer (ABS), 22:984
    thiols in, 24:29
Acrylonitrile-butadiene-styrene
        terpolymer
    silane coupling agent for, 22:152

Acrylonitrile-styrene copolymer, 22:1023
6-O-Acryloylsucrose, 23:75
Actilight, 23:74
Actinobacillosis, 24:829
Actinomycosis, 24:829
Activated carbon [7440-44-0]
    regeneration using SCFs, 23:468
    in silver processing, 22:170
    in sodium bromide prdn, 22:378
    water treatment for steam prdn, 22:745
Activation analysis, 22:656
Activators
    rubber chemicals, 21:460
Activity coefficient, 23:1015
Actuators
    shape-memory, 21:970
Acylation
    sulfonic acids as catalyst, 23:209
Acylthiophenes, 24:43
Addition-curing sealants, 21:654
Addition products
    from sulfur monochloride, 23:286
Additives
    in styrene plastics, 22:1058
Adenosine [58-61-7], 22:936
Adenylyl cyclase, 23:578
Adhesion
    of thin films, 23:1074
    work of, 21:211
Adhesion promoters
    rubber chemicals, 21:460
    from silanes, 21:659; 22:48
Adhesives
    as paper contaminants, 21:11
    PVA resins as, 24:980
    PVB in, 24:935
    rheological measurements, 21:427
    as roofing material, 21:440
    SBR use, 22:1011
    as sealants, 21:651
    silica gels for, 21:1000
    silicates as, 22:23
    sorbitol in, 23:112
    starch in, 22:712
    thioglycolic acid as, 24:13
    titanates as, 24:290, 323
Adios
    insect repellent, 21:242
Adjuvants, 24:742
Adogen
    surfactant, 23:529
Adsorbents
    regeneration, 23:465

regeneration using SCFs, 23:468
silica gel as, 21:1001, 1022
silica sols as, 21:1000
silicates as, 22:25
for sodium chloride, 22:368
Adsorption
of hydrogen sulfide, 23:433
on silicate surfaces, 22:11
Advanced ceramics
as tool materials, 24:428
Advanced wastewater treatment (AWT),
21:165
Aerogels, 21:996, 1020; 22:25, 499
reactions using SCF, 23:465
silicon esters in prprn of, 22:79
by sol–gel technology, 22:500
supercritical drying of, 23:473
Aerosil, 21:998
Aerosols
analysis of heavy metals in, 22:646
trace analysis of, 24:518
Aerospace shape-memory alloys, 21:968
Aerotil L Soil Conditioner, 22:460
Aflatoxins, 22:600
Agglomeration, 22:223
Aging
role in sol–gel technology, 22:508
Agitation, 22:229
Agrichemicals
thiols in, 24:29
Air
tanks for, 23:644
use in optical spectroscopy, 22:645
Air-assisted atomizer, 22:672
Airblast atomizer, 22:672
Air classifiers, 22:291
Air-conditioning systems
sulfamic acid cleaner for, 23:129
Aircraft sealants, 21:665
Aircraft structures
brazing filler metals for, 22:490
Air pollution control, 21:156
Air quality control region (AQCR), 21:165
Alabandite
magnetic intensity, 21:876
Alachlor [15972-60-8], 22:422, 425, 427
Alanine
in silk, 22:156
in soybeans and other oilseeds, 22:596
Albendazole [54965-21-8], 24:831, 832
Alberger salt, 22:363
Alcaligenes faecalis
inhibited by sorbates, 22:580

Alcogel
by sol–gel technology, 22:500
Alcohol. See Ethanol.
Alcohol dehydrogenase
in teas, 23:751
Alcoholic fermentation, 24:841
Alcohols
reactions with silanes, 22:51
trace analysis of, 24:503
Alcohol sulfates, 23:170
Alcosorb, 22:460
Aldehyde [37677-14-8], 23:849
Aldicarb [116-06-3], 22:423
Alditols, 23:93
Aldo
glycerol ester surfactant, 23:512
Aldo PMS
fatty acid surfactant, 23:518
Aldoses
from sugar alcohols, 23:102
Aldosterone [6251-69-0], 22:861
Aldrin [309-00-2], 22:421
Aleuritic acid
in shellac, 21:300
Alfonic, 23:502
alcohol surfactant, 23:502
ethoxylated alcohol surfactant, 23:508
Alitame [80863-62-3], 23:561, 569
sucrose contrasted, 23:4
Alkali–gravity–viscosity charts, 22:13
Alkaloids, 24:506
Alkamide Rodea
diethanolamine surfactant, 23:520
Alkamuls
castor oil surfactant, 23:517
Alkane
sulfurization of, 23:429
Alkanolamines
for hydrogen sulfide removal, 23:436
Alkar process, 22:967
Alkoxides
of titanium, 24:258, 275
Alkoxysilanes, 22:70
Alkyd resins
for soybeans and other oilseeds, 22:611
use in refractory brick, 21:69
Alkyds
resin formation, 24:326
Alkylarenesulfonates, 23:495
Alkylation
by dialkyl sulfates, 23:412
for styrene, 22:966
sulfonic acids as catalyst, 23:209

Alkylbenzenesulfonates, 23:494
    sampling standards for, 21:627
Alkylbenzenesulfonate surfactant, 23:312
Alkylene ureas
    in textile finishing, 23:899
Alkylphenol disulfide [68555-98-6]
    cross-linking agent, 21:470
Alkylphenol ethoxylates, 23:510
Alkyl polyglycosides, 23:518
Alkylthiophenes, 24:34
Allergic reactions
    to latex products, 21:585
Allergies
    vaccines against, 24:743
Allitol [488-44-8], 23:95
4-trans-6-cis-Alloocimene [7216-56-0],
    23:849
4-trans-6-trans-Alloocimene [3016-19-1],
    23:849
Alloocimene diepoxide [3765-28-4], 23:850
Alloocimene isomers [637-84-7], 23:850
Alloocimenols [18479-54-4], 23:850
Allotelluric acid [13520-55-3], 23:798
Alloy 20, 23:393
Alloy steels, 22:816
Allyl alcohol [107-18-6]
    as solvent, 22:536
4-Allylanisole [140-67-0]
    as insect repellent, 21:253
Allyl chloride [107-05-1], 24:86
Allyl vinyl sulfoxide [81898-53-5], 23:217
Almonds
    oil removal from, 23:467
Almond shells, 23:96
Alonized, 23:390
Alpha-Step MC48, 23:500
Alrowet
    dialkyl sulfosuccinate surfactant,
    23:498
Altaite [12037-86-4]
    tellurium in, 23:783
Alternaria citri
    inhibited by sorbates, 22:579
Alternaria spp.
    inhibited by sorbates, 22:579
Alternaria tenuis
    inhibited by sorbates, 22:579
Altritol [5552-13-6], 23:95
D-Altritol [17308-29-1], 23:95
D,L-Altritol [60660-57-3], 23:95
L-Altritol [60660-58-4], 23:95
Alumina [1344-28-1], 24:606, 854
    refractories from, 21:50

    as silica coating, 21:1020
    sol–gel processing, 22:519
    solubility in steam, 22:729
    support for enzymes, 23:595
    thickening of, 21:680
    thin films of, 23:1061
    as tool materials, 24:427
Alumina, activated
    sodium on surface, 22:328
Alumina beads, 23:446
Alumina hydrate
    thickening of, 21:680
α-Alumina trihydrate [14762-49-3]
    in rubber, 21:480
Alumina–zirconia alloy
    as tool material, 24:431
Aluminides
    as refractory coatings, 21:103
Aluminosilicate clays, 23:734
Aluminosilicate fibers, 21:127
    as refractory coatings, 21:102
Aluminum, 23:1045
    brazing of, 22:484
    corrosion rates in acid, 23:122
    critical surface tension of, 22:148
    diffusion coefficient in vitreous silica,
        21:1047
    electroplated thin films, 23:1072
    grain refining agent, 24:256
    role in steelmaking, 22:779
    as sampling probe, 21:634
    sampling standards for, 21:627
    in scrap for steel, 22:779
    separation by reverse osmosis, 21:322
    in silicon and silica alloys, 21:1106
    sulfur dioxide absorption, 23:447
    for tanks and pressure vessels, 23:644
    in tea, 23:751
    thin films of, 23:1060
    tool materials for, 24:436
    vapor pressure of, 23:1046
    vitreous silica impurity, 21:1036
Aluminum alkoxide, 22:520
Aluminum alloys
    brazing filler metals, 22:491
    dielectric constants of, 22:761
    grain refining agent, 24:256
    as shape-memory alloys, 21:964
    silicon in, 21:1109
    trace analysis of, 24:518
    use in sulfuric acid manufacture, 23:390

Aluminum antimonide [25152-52-7], 21:767

Aluminum arsenide [22831-42-1], 21:767

Aluminum butoxide
use in sol–gel technology, 22:526

Aluminum–copper alloys
sodium for, 22:347

Aluminum furnaces
refractories for, 21:83

Aluminum hydroxide
centrifugal separation of, 21:850

Aluminum nitride [24304-00-5]
in steels, 22:808

Aluminum oxide [1344-28-1], 21:119, 847; 22:519
as refractory material, 21:57
substrate for self-assembled monolayer, 23:1089

Aluminum phosphide [20859-73-8], 21:767

Aluminum salicylate [18921-11-4], 21:612

Aluminum–silicon alloys
strontium as eutectic modifier, 22:949
tool materials for, 24:436

Aluminum soap, 22:324

Aluminum sulfate
use of sulfuric acid, 23:397

Aluminum telluride [12043-29-7], 23:793

Aluminum titanate, 24:253

Alvar, 24:925

Amalgam process
for prdn of sodium dithionite, 23:317

Amaranth starch, 22:712

Amber, 21:298

Ambient air, nonpolluted
in vacuum systems, 24:759

American Welding Society
classification of brazing filler metals, 22:487

Amersil, 21:1049

Ames test
for thioglycolic acid, 24:11

Amidosulfonates, 23:498

Amidosulfuric acid, 23:120

Amidox
polyoxyethylene surfactant, 23:521

Amiloride [2016-88-8], 23:578

Aminations
reactions using SCF, 23:465

Amines, 22:730
reactions with silanes, 22:52
silver complexes of, 22:185
thorium complexes, 24:74, 75
water additive of steam prdn, 22:740

Amine soaps, 22:324

Amino acids
chromatography of, 22:145
dextrose in prdn of, 23:593
in silk, 22:156
in tea, 23:750

p-Aminobenzenesulfonamide [63-74-1], 24:828

m-Aminobenzenesulfonic acid [121-47-1], 23:203

3-Amino-2-(4-chlorophenyl)propane-sulfonic acid, 23:211

2-Aminoethanesulfonic acid, 23:210

N-(2-Aminoethyl)-3-aminopropyl-trimethoxysilane [1760-24-3]
adhesion promoter, 21:656
critical surface tension of, 22:148
silane coupling agent, 22:150

Aminoglycosides
as veterinary drugs, 24:828

(17-Aminoheptadecyl)-trimethoxysilane
self-assembled monolayers of, 23:1090

α-Amino-3-hydroxy-5-methylisoxazole-4-propionate, 22:935

Aminol
diethanolamine surfactant, 23:520

Aminolauric acid, 22:526

6-Aminopenicillanic acid [551-16-6], 24:828
silylation of, 22:146

4-[(4-Aminophenyl)(4-imino-2,5-cyclohexadien-1-ylidene)methyl]-N-phenyl aniline, monohydrochloride [68966-31-4], 24:554

Aminophylline [317-34-0], 22:936

Aminoplasts
in grouting systems, 22:457

Aminopropyltriethoxysilane
critical surface tension of, 22:148

3-Aminopropyltriethoxysilane [919-30-2]
adhesion promoter, 21:656
silane coupling agent, 22:150

3-Aminopropyltrimethoxysilane [13822-56-5]
adhesion promoter, 21:656

4-Aminopyridine [504-24-5]
as bird repellent, 21:255

Amino resins
use in refractory brick, 21:69

p-Aminosalicylic acid [65-49-6], 21:616

Amitriptyline [50-48-6], 22:941

Ammo hypo, 24:61

Ammonia [7664-41-7]
  boiling point of, 23:627
  as catalyst, 22:499
  reactions with silanes, 22:52
  as refrigerant, 21:131, 134, 138
  as rubber latex preservative, 21:581
  scrubbing, 23:379
  sensors for, 21:823
  separation by reverse osmosis, 21:323
  silver complexes of, 22:185
  sodium solutions in, 22:329
  solubility in steam, 22:730
  specific gravity, 23:625
  in steam, 22:729
  steam in prdn of, 22:757
  supercritical fluid, 23:454
  tanks and pressure vessels for, 23:631
  water additive of steam prdn, 22:740
Ammonia mercerization, 23:894
Ammonium N-benzoylsulfamate [83930-
    12-5], 23:124
Ammonium chloride [12125-02-9], 24:856
  solubility in steam, 22:729
Ammonium dithiocarbamate [513-74-6],
    23:321
Ammonium diuranate [7783-22-4], 24:650
Ammonium hexachlorotitanate [21439-
    26-9], 24:261
Ammonium metavanadate [7803-55-6],
    24:800
Ammonium perrhenate [13598-65-7],
    21:337
Ammonium persulfate [7727-54-0]
  as grouting catalyst, 22:455
Ammonium salicylate [528-94-9], 21:612
Ammonium sulfamate [7773-06-0], 23:125
Ammonium sulfate
  centrifugal separation of, 21:870
Ammonium sulfide [12135-76-1], 23:269
Ammonium sulfite [10192-30-0], 23:301
Ammonium tetrachloroferrate [24411-
    12-9], 24:858
Ammonium tetrathiotungstate [13862-
    78-7], 24:597
Ammonium thiocyanate [1762-95-4],
    23:320, 321
Ammonium thioglycolate [5421-46-5], 24:8
Ammonium thiosulfate [7783-18-8], 24:55
Ammonium tungstate [11140-77-5], 24:593
Ammonix LO
  amine oxide surfactant, 23:525
Ammonix MO
  amine oxide surfactant, 23:525

Ammonyx CO
  amine oxide surfactant, 23:525
Ammonyx DMCD-40
  amine oxide surfactant, 23:525
Ammonyx MO
  amine oxide surfactant, 23:525
Ammonyx SO
  amine oxide surfactant, 23:525
Amorphous semiconductors, 21:750
Amorphous silica, 21:1005
Amoxapine [14028-44-5], 22:942
Amperometric cells
  as sensors, 21:825
Amphetamine [300-62-9], 22:937
Amphiboles, 21:980
Amphosol
  surfactant, 23:531
Amphotericin B [1397-89-3], 24:829
Amprolium [121-25-5], 24:831
AMPS monomer, 23:208
Amyl acetate [628-63-7]
  as solvent, 22:542
α-Amylase
  for dextrose prdn, 23:586
Amylopectin [9037-22-3], 22:702
Amylose [9005-82-7], 22:702
Amyl salicylate [2050-08-0], 21:614
η-Amyrin [559-70-6], 22:898
Analeptics, 22:932
Analgesics
  salicylic acid derivatives, 21:602
  as veterinary drugs, 24:835
Anatase [1317-70-0], 24:235
Andalusite [12183-80-1], 21:54
Andersonite [12202-87-8], 24:643
Androgen, 22:852
Androsta-1,4-diene-3,17-dione [897-06-3],
    22:876
Androsta-4,9-diene-3,17-dione [1035-69-4],
    22:878
Androstane [24887-75-0], 22:854
Androst-4-ene-3,17-dione [63-05-8], 22:878
Androsterone [53-41-8], 22:859
Anesthetics
  steroids as, 22:852
  as veterinary drugs, 24:834
Anethole [104-46-1], 23:836
ANFO
  use in salt mining, 22:364
Angioplasty
  shape-memory alloy devices for, 21:974
Anhydrite
  source of sulfur, 23:246

Anhydrization
  of sugar alcohols, 23:100
Anhydrosorbitol esters, 23:516
Anhydrothiosulfuric acid [83682-21-7],
    24:54
Aniline
  boiling point of, 23:627
Aniline cloud point, 22:535
8-Anilino-1-naphthalenesulfonic acid
    [82-76-8], 23:211
Animal drug controls, 21:177
Animal fat
  sulfurization of, 23:429
Animal feeds
  use of sorbates, 22:584
Animal waxes
  as release agents, 21:210
Anisole [100-66-3], 24:856
Ankerite
  magnetic intensity, 21:876
meta-Ankoleite [12169-00-5], 24:643
Annealing
  of steel, 22:805
Anodes
  tin, 24:115
Anodized aluminium
  triarylmethane dyes for, 24:565
Antarox
  ethoxylated alcohol surfactant, 23:508
Anthelmintics
  as veterinary drugs, 24:830
Anthracene
  analysis using laser-induced
      fluorescence, 22:654
  rejection by reverse osmosis membrane,
    21:319
Anthrahydroquinones
  as bird repellents, 21:256
Anthranilic acid [118-92-3], 23:566
Anthraquinone [84-65-1]
  as bird repellent, 21:256
Anthraquinone disulfonic acid
  use in sulfur recovery, 23:444
Anthraquinone sulfonic acid, 23:158
2-(9-Anthryl)ethyl chloroformate, 24:506
Antiasthmatics
  steroids as, 22:852
Antibiotics, 24:727
  centrifugal separation of, 21:861
  dextrose in prdn of, 23:592
  steroids as, 22:852
  thiophene and thiophene derivatives
      as, 24:48

Antiblocking agents, 21:207
Antibodies, 24:508
  supercritical fluid extraction, 23:465
Anticaking agents
  silica as, 21:1001
  for sodium chloride, 22:368
Anticancer agents
  steroids as, 22:852
Antidegradants, 21:516
  for rubber, 21:574
  rubber chemicals, 21:460
Antidepressants, 22:938
Antifoams
  in adhesives, 24:971
  silicas in, 21:1002
  water additive of steam prdn, 22:740
Antifog agents, 23:112
Antihormones
  steroids as, 22:852
Antiinfectants
  sodium iodide as, 22:382
Antiinflammatory agents
  salicylic acid derivatives, 21:602
  steroids as, 22:852
  thiophene and thiophene derivatives,
    24:47
  as veterinary drugs, 24:832
Anti-Markovnikoff addition, 24:23
Antimicrobial agents
  in soap, 22:320
  as veterinary drugs, 24:827
Antimicrobial effectiveness
  of sorbates, 22:584
Antimony, 23:1046
  dopant for silicon, 21:1098
  in scrap for steel, 22:779
  silver cmpds in detm of, 22:190
  in solders, 22:485
  in titanium pigments, 24:248
  vitreous silica impurity, 21:1036
Antimony oxide [1309-64-4]
  in rubber, 21:480
Antimony pentafluoride [7783-70-2]
  plus triflic acid, 23:209
Antimony–tin blue, 24:127
Antimony trifluoride, 24:288
Antimony tris(2-ethylhexyl thioglycolate)
    [26888-44-4], 24:16
Antimycotics
  as veterinary drugs, 24:829
Antioxidants, 23:430; 24:14
  in lubricating oils and greases, 23:804

for rubber, *21*:574
selenium as, *21*:713
in soap, *22*:320
sodium nitrite use in, *22*:400
in styrene plastics, *22*:1058
sulfur compounds as, *23*:291
in teas, *23*:762
Antiozonants
sodium nitrite use in, *22*:400
Antiparasitic agents
as veterinary drugs, *24*:830
Antiperspirants
soaps in, *22*:324
Antiprotozoals
as veterinary drugs, *24*:830
Antipyretics
salicylic acid derivatives, *21*:602
Antiseptics
silver cmpds as, *22*:192
Antistatic agents
in styrene plastics, *22*:1058
for textiles, *23*:112
Antistick agents, *21*:207
Anti-Stokes lines, *22*:649
Antiviral agents
sulfonic derivatives as, *23*:211
Antoine equation, *23*:990
for vanadium, *24*:784
Aosoft, *24*:1079
Apatite
magnetic intensity, *21*:876
Aphthitalite, *22*:404
APM, *23*:558
Apollo, *23*:1036
vitreous silica windows, *21*:1068
Apophyllite [*1306-03-2*], *21*:980
Apples, *23*:96
Apricots, *23*:96
AQCR. See *Air quality control region.*
Aquaculture, *23*:980
Aquaflex, *24*:1079
Aquasorb, *22*:460
Aquastore, *22*:460
Arabinitol [*2152-56-9*], *23*:95
D-Arabinitol [*488-82-4*], *23*:95
D,L-Arabinitol [*6018-27-5*], *23*:95
L-Arabinitol [*7643-75-6*], *23*:95
Arachidic acid
thin films of, *23*:1081
Arachin
in peanuts, *22*:595
Aramid
in tire cord, *24*:163

Argentite [*1332-04-3*], *22*:170, 179
Arginine
in silk, *22*:156
in soybeans and other oilseeds, *22*:596
Argon
diffusion in vitreous silica, *21*:1048
removal from natural gas, *23*:139
role in steelmaking, *22*:781
for sputter deposition, *23*:1050
Argon–oxygen decarborizers
refractories for, *21*:83
Argyria, *22*:188
Argyrosis, *22*:188
Armotan
sorbitan surfactant, *23*:516
Armotan ML
surfactant, *23*:515
Armotan MO
surfactant, *23*:515
Armotan MS
surfactant, *23*:515
Aromatic 100 solvent, *22*:540
Aromatic 150 solvent, *22*:540
Aromox C/12
amine oxide surfactant, *23*:525
Aromox DM16
amine oxide surfactant, *23*:525
Aromox DMCD
amine oxide surfactant, *23*:525
Aromox DMHTD
amine oxide surfactant, *23*:525
Aromox T/12
amine oxide surfactant, *23*:525
Arosurf
ethoxylated alcohol surfactant, *23*:508
Arprinocid [*55779-18-5*], *24*:831
Arquad
surfactant, *23*:529
Arrhenius rate theory, *22*:834
Arrowroot
dextrose from, *23*:585
Arsenic [*7440-38-2*]
dopant for silicon, *21*:1098
impurity in selenium, *21*:705
impurity in vitreous silica, *21*:1036
regulatory level, *21*:160
silver cmpds in detm of, *22*:190
in sulfuric acid, *23*:399
Arsenopyrite [*1303-18-0*]
sulfide ore, *23*:244
Arthritis
treatment using thiophene, *24*:47
veterinary drugs against, *24*:832

Asbestos
    sampling standards for, *21*:627
Asbestos regulations
    in power plants, *21*:193
Ascochyta cucumis
    inhibited by sorbates, *22*:579
Ascochyta sp.
    inhibited by sorbates, *22*:579
Ascorbic acid
    oxygen scavenger for steam, *22*:742
    in steam, *22*:737
Aspartame [*22839-47-0*]
    compared to sucrose, *23*:4
    as sweetener, *23*:558
Aspartic acid
    in soybeans and other oilseeds, *22*:596
Aspartic acid–asparagine
    in silk, *22*:156
L-Aspartyl-L-phenylalanine methyl ester,
    23:558
Aspergillus clavatus
    inhibited by sorbates, *22*:579
Aspergillus elegans
    inhibited by sorbates, *22*:579
Aspergillus flavus, *22*:600
    inhibited by sorbates, *22*:579
Aspergillus fumigatus
    inhibited by sorbates, *22*:579
Aspergillus glaucus
    inhibited by sorbates, *22*:579
Aspergillus niger
    inhibited by sorbates, *22*:579
Aspergillus ocraceus
    inhibited by sorbates, *22*:579
Aspergillus parasiticus
    inhibited by sorbates, *22*:579
Aspergillus sydowi
    inhibited by sorbates, *22*:579
Aspergillus terreus
    inhibited by sorbates, *22*:579
Aspergillus unguis
    inhibited by sorbates, *22*:579
Aspergillus versicolor
    inhibited by sorbates, *22*:579
Asphalt, *23*:719
    as roofing material, *21*:439
    sulfur-extended binders, *23*:261
    tanks for, *23*:625, 647
Asphaltenes, *23*:719
Asphaltites, *21*:299
Asphalt–rubber compounds
    from scrap tires, *21*:31

Aspirin [*50-78-2*], *21*:610, 617; *24*:832
Assaying triarylmethane dyes
    Knecht method, *24*:556
Astrakanite, *22*:386, 404
Asymmetric synthesis
    titanium catalysts for, *24*:308
Athabasca sand, *23*:719
Atmospheric drying
    of soap, *22*:315
Atomic absorption
    for trace analysis, *24*:498
Atomic absorption spectroscopy, *22*:645
Atomic emission spectroscopy, *22*:645
Atomic fluorescence spectroscopy, *22*:653
Atomic force microscopy, *23*:773
    for self-assembled monolayers, *23*:1091
Atomization, *22*:245
Atomizers, *22*:670
Atrazine [*1912-24-9*], *22*:422, 427; *24*:510
Atropine, *24*:830
Attemperation
    in steam prdn, *22*:745
Attention-deficit hyperactivity disorder
    stimulants for, *22*:937
Auger spectroscopy, *21*:90, 110
Auramine base [*492-80-8*], *24*:567
Auramine G [*2151-60-2*], *24*:566
Auramine O [*2465-27-2*], *24*:566
Auramine sulfate [*52497-46-8*], *24*:567
Aurora
    vitreous silica optics for, *21*:1066
Austempering
    of steel, *22*:804
Austenite, *22*:791
Autoclave molding, *21*:197
Autoclaves
    for sterilization, *22*:839
Autoclave tape, *22*:843
Autoimmune diseases
    vaccines against, *24*:743
Automobiles
    shape-memory actuators for, *21*:970
Automobile window interlayers, *24*:932
Automotive industry
    steel for, *22*:822
Automotive parts
    brazing filler metals for, *22*:490
Autovibron, *21*:420
Autunite [*16390-74-2*], *24*:643
Avalanche photodiodes, *21*:798
Δ^5-Avenasterol
    in soybeans and other oilseeds, *22*:597

Avermectins
  as veterinary drugs, *24*:830
Avicennite [*12022-82-1*], *23*:952
Avirol
  alcohol surfactant, *23*:502
Avirol SA 4106
  surfactant, *23*:501
Avirol SA 4110
  surfactant, *23*:501
Avirol SL 2010
  surfactant, *23*:501
Avitrol [*504-24-5*]
  as bird repellent, *21*:255
Avocado
  sugar alcohols in, *23*:96
Avon Skin-So-Soft
  as mosquito repellent, *21*:244
Avoparcin [*37332-99-3*], *24*:829
AWT. See *Advanced wastewater
    treatment.*
4'-Azido-4'-deoxy-β-D-sorbofuranoside,
    23:71
4'-Azido-4'-deoxysucrose, *23*:71
2,2'-Azinobis-(3-ethyl benzothiazoline-
    sulfonate), *23*:763
Azodicarbonamide [*123-77-3*]
  as blowing agent, *21*:479
Azo dyes
  sulfonic acid-based, *23*:206
Azo-stilbene dyes, *22*:923
Azotobacter agilis
  inhibited by sorbates, *22*:580

B

Babassu oil
  in soap manufacture, *22*:302
Babbitt, *24*:119
Babbit-type alloys
  tellurium in, *23*:803
Baccilus typhosus
  effect of silver, *22*:174
Bacdanol [*28219-61-6*], *23*:862
Bacillus cereus
  inhibited by sorbates, *22*:580
Bacillus coagulans
  inhibited by sorbates, *22*:580
Bacillus polymyxa
  inhibited by sorbates, *22*:580
Bacillus stearothermophilus
  inhibited by sorbates, *22*:580

Bacillus subtilis
  inhibited by sorbates, *22*:580
Bacitracins
  as veterinary drugs, *24*:829
BACT. See *Best available control
    technology.*
Bacteria
  centrifugal separation of, *21*:860
  inhibited by sorbates, *22*:580
  preventors in cooling towers, *22*:217
Bacterins
  as veterinary drugs, *24*:837
Bacteriostasis, *22*:848
Baddeleyite [*12036-23-6*], *21*:54
  refractories from, *21*:50
Bagasse, *23*:22
  fuel for steam prdn, *22*:746
Bagasse fiber, *23*:42
Bahnmetall, *22*:347
Bainite, *22*:791
  phase properties of, *22*:798
Baiyunoside [*86450-75-1*]
  potential sweetener, *23*:573
Baked goods
  sucrose in, *23*:6
Bakery products
  dextrose for, *23*:593
  use in sorbates, *22*:583
Baking
  high fructose syrup in, *23*:597
  vanillin in, *24*:819
Balances
  silica in, *21*:1000
Balling, *22*:231
Ball mills, *22*:287
Balm of Gilead [*8022-26-2*], *21*:298
Balsamic vinegar
  prdn process, *24*:843
Balsam Mecca [*8022-26-2*], *21*:298
Balsam of Peru [*8007-00-9*], *21*:298
Balsam of Tolu [*9000-64-0*], *21*:298; *24*:350
Banana starch, *22*:712
Band theory
  for semiconductors, *21*:722
Barbiturates
  as veterinary drugs, *24*:835
Barium [*7440-39-3*]
  complexes with sugar alcohols, *23*:103
  in magnesium ferrosilicon, *21*:1118
  regulatory level, *21*:160
Barium carbonate
  sodium sulfide as by-product, *22*:416

Barium hexafluorotitanate [31252-69-6], 24:256

Barium selenate [7787-41-9], 21:698

Barium selenite [13718-59-7], 21:698

Barium–silicon alloy, 21:1119

Barium–sodium alloys, 22:347

Barium stannate [12009-18-6], 24:128

Barium sulfonates, 23:212

Barium thiosulfate monohydrate [7787-40-8], 24:53

Barium titanate [12047-27-7], 24:252, 301

Barlox 12
    amine oxide surfactant, 23:525

Barlox 14
    amine oxide surfactant, 23:525

Barlox 10S
    amine oxide surfactant, 23:525

Barquat
    surfactant, 23:529

Barrier films, 23:1049; 24:902

Barrier latex, 24:916

Barrier properties
    of vinylidene chloride polymers, 24:898

Barry arylation, 22:57

Bar soap, 22:313

Basalts
    thorium dating of, 24:70

Basic Blue 26 [2580-56-5], 24:560

Basic Green 4 [569-64-2], 24:553

Basic-oxygen furnaces
    refractories for, 21:47

Basic-oxygen process
    role in steelmaking, 22:773

Basic Red 18 [14097-03-1], 24:552

Basic tin citrate [59178-29-9], 24:122

Basic Violet 1 [8004-87-3], 24:562

Basic Violet 2 [3248-91-7], 24:561

Basic Violet 3 [548-62-9], 24:552

Basic Violet 4 [2390-59-2], 24:560

Basic Yellow 37 [6358-36-7], 24:566

Basil
    extraction using SCFs, 23:466

Basogel, 22:79

Bastnasite
    magnetic intensity, 21:876

BAT. See Best available technology.

Batteries, 23:140
    cardiac pacemaker, 23:1036
    silica in, 21:1002
    sodium for, 22:345
    solar energy charging, 22:475
    strontium in, 22:950
    titanium sulfides in, 24:266

Batteries, storage
    tellurium in, 23:803

Batteries, primary
    silver cmpds in, 22:190

Battery chargers
    solar energy for, 22:475

Battery grids
    use of selenium in, 21:711

Baumé scale
    for sulfuric acid, 23:368

Bauxite [1318-16-7], 21:53, 681
    for hydrogen sulfide prdn, 23:281
    refractories from, 21:49

Bayer process, 21:54, 681

Bayleyite [19530-04-2], 24:643

BCT. See Best conventional pollutant control technology.

Beach sands
    cleaning by magnetic separation, 21:895

Beam spectroscopy, 22:657

Bearing alloys
    tellurium in, 23:803

Bearing materials
    sodium in, 22:347
    tin alloys as, 24:119
    tools for, 24:445

Bearings
    thallium alloys in, 23:953
    use of tellurium in, 23:802

Beavon process, 23:268, 444

Beclomethasone dipropionate [55340-19-8], 22:905

Becquerelite [12378-67-5], 24:643

Beer, 21:1020; 23:98
    dextrose in, 23:592
    silica gel in prdn of, 21:1000
    use of sulfur dioxide in mfg, 23:310

Beer's law, 22:628

Beet molasses, 23:58

Beet pulp
    use of sorbates, 22:584

Beets
    sulfoxides in, 23:217

Beet sugar, 23:44

Belts
    precipitated silica in, 21:1025

Bendiocarb [22781-23-3]
    as insect repellent, 21:253

Benefield, 23:439

Benitoite [15491-35-7], 22:3; 24:264

Bentazon [25057-89-0], 22:425; 23:295

Benthiocarb [28249-77-6], 23:271

Bentonite [1302-78-9], 21:980
  membrane for ultrafiltration, 24:606
Benzaldehyde, 24:386
Benzamide, 24:503
Benzene [71-43-2], 24:856
  boiling point of, 23:627
  dielectric constants of, 22:759
  as HAP compound, 22:532
  regulatory level, 21:160
  rejection by reverse osmosis membrane,
    21:319
  as solvent, 22:540
  specific gravity, 23:625
  sulfonation of, 23:157
Benzenedisulfonic acid (disodium salt)
    [831-59-4], 23:206
Benzenesulfonic acid [98-11-3], 23:157,
    194
Benzenetellurol [69577-06-6], 23:800
Benzenethiol [108-98-5], 24:21
Benzimidazoles
  as veterinary drugs, 24:831
Benzo[a]pyrene
  analysis using laser-induced
    fluorescence, 22:654
Benzo[b]selenophene [272-30-0], 21:703
Benzoic acid [65-85-0]
  as rubber chemical, 21:473
Benzoselenazole [273-91-6], 21:714
1,2,3-Benzotriazole
  determination of silver, 22:187
Benzoyl peroxide [94-36-0]
  cross-linking agent, 21:470
  silicone peroxide cure, 22:108
Benzyl acetate [140-11-4]
  as solvent, 22:542
Benzyl alcohol [100-51-6]
  as solvent, 22:536
Benzyl benzoate [120-51-4], 21:240
Benzyl chloride
  from toluene, 24:384
4,6-O-Benzylidenesucrose, 23:65
Benzyl salicylate [118-58-1], 21:614
Berberine, 24:506
Berlin-Blue
  for thallium poisoning, 23:961
Berlin Institute method
  reducing sugar test, 23:16
Berries
  dextrose in, 23:583
Bertholet's silver, 22:183
Beryl [1302-52-9], 21:980

Beryllia, 21:56
Beryllides
  as refractory coatings, 21:103, 106
Beryllium
  in brazing filler metals, 22:492
Beryllium oxide [1304-56-9]
  as refractory material, 21:57
  solubility in steam, 22:728
Bessemer process
  for steel, 22:766
Best available control technology (BACT),
    21:165
Best available technology (BAT), 21:165
Best conventional pollutant control
    technology (BCT), 21:165
Best practicable control technology
    (BPCT), 21:165
Best R&D practices, 21:270
Betaine
  in beet sugar, 23:55
Betamethasone, 24:832
Beverages
  dextrose for, 23:593
  high fructose syrup in, 23:597
  sugar alcohols in, 23:109
Beverages, carbonated. See Carbonated
    beverages.
BHC [58-89-9], 22:421
Bicuculline [485-49-4], 22:934
BiDMC. See Bismuth
    dimethyldithiocarbamate.
Bile acids, 22:852
Bill of lading, 24:531
Binders
  ethoxysilanes as, 22:78
  silica sols as, 21:1000
  sulfur in, 23:261
Binding agents
  titanate chelates as, 24:291
Binding mechanisms, 22:224
Bingham fluids, 21:366
Biochemical oxygen demand, 21:165
Biochemical systems
  analysis using Raman spectroscopy,
    22:650
Biocides, 23:976
  in adhesives, 24:971
  in grouts, 22:456
  PVC as, 24:1036
  sodium bromide as, 22:379
Biocontamination
  of latex, 24:966

Biocytin, *24*:504
Biodegradability
  of sulfonates, *23*:162
Biodegradation
  of pesticide, *22*:432
  of polystyrene, *22*:1033
  of poly(vinyl alcohol), *24*:993
Bio-FGD process, *23*:449
Biofuels, *21*:226
Biological products
  regulatory agencies, *21*:174
Biomass
  from solar energy, *22*:476
Biomass fuel, *21*:226
Biomaterials
  silk as, *22*:162
Biomedical devices
  shape-memory alloys for, *21*:973
  from sol–gel technology, *22*:526
  use of tantalum in, *23*:671
Biomedical diagnostics
  optical spectroscopic analysis, *22*:644
Bioremediation
  use of selenium, *21*:714
Bio-Soft
  alkylbenzensulfonate surfactant,
    23:496
  ethoxylated alcohol surfactant, *23*:508
Bio-SR process, *23*:445
Biosyn, *23*:543
Biotechnology, *24*:516
  pest control programs, *22*:447
Bio-Terge
  α-olefinsulfonate surfactant, *23*:497
Biotin ethylenediamine, *24*:504
6-(Biotinoylamino)caproic acid, *24*:504
6-(Biotinoylamino)caproic acid hydrazide,
    24:504
Biotite
  magnetic intensity, *21*:876
Bipolar junction transistor (BJT), *21*:782
Bipolar structure
  of soap, *22*:298
Birch
  xylose from, *23*:105
Bird feathers
  trace analysis of, *24*:518
Bird repellents, *21*:254
Bis(2,2′-benzothiazolyl)disulfide [*120-78-
    5*], *21*:465
2,2′-Bis(*t*-butylperoxy)diisopropylbenzene
    [*25155-25-3*]
  cross-linking agent, *21*:470

1,1-Bis(*t*-butylperoxy)-3,3,5-trimethyl
    cyclohexane [*6731-36-8*]
  cross-linking agent, *21*:470
Bis(4-chlorophenyl) sulfone [*80-07-9*],
    23:141
  from thionyl chloride, *23*:295
4,4′-Bis(*p*-chlorophenylsulfonyl) biphenyl
    [*22287-56-5*], *23*:141
Biscyclopentadienyls
  titanium compounds, *24*:314
Bis(cyclopentadienyl)titanium dichloride
    [*1271-19-8*], *24*:275, 307
Biscyclopentadienylvanadium chloride
    [*12083-48-6*], *24*:800
4,4′-Bisdiazo-2,2′-stilbenedisulfonic acid
    [*57153-16-9*], *22*:925
4,4′-Bis(dimethylamino)benzophenone
    (Michler's ketone) [*90-94-8*], *24*:560
Bis(dimethylamino ethyl ether), *24*:699
1,3-Bis(ditraconimidomethyl)benzene
    [*119462-56-5*]
  cross-linking agent, *21*:471
Bisflavanols, *23*:756
Bis(2-hydroxy-3,5,6-trichlorophenyl)-
    methane, *22*:848
1,3-Bis(isocyanatomethyl)cyclohexane
    [*38661-72-2*], *24*:708
Bis(4-methoxyphenyl) sulfoxide [*1774-
    36-3*], *23*:225
Bismuth
  in brazing filler metals, *22*:492
  in solders, *22*:485
Bismuth antimony telluride
  thermoelectric material, *23*:1034
Bismuth dimethyldithiocarbamate
    (BiDMC) [*21260-46-8*], *21*:465
Bismuth salicylate [*5798-98-1*], *21*:612
Bismuth stannate [*12777-45-6*], *24*:128
Bismuth subsalicylate [*5798-98-1*], *21*:612
Bismuth telluride [*1304-82-1*], *23*:791
  thermoelectric material, *23*:1034
Bisphenol A
  thioglycolic acid for mfg of, *24*:15
Bisphenol S, *23*:141
N-(4-(Bis[4-(phenylamino)phenyl]-
    methylene)-2,5-cyclohexadien-1-
    ylidene)-3-methyl-benzeneamine
    sulfate [*57877-94-8*], *24*:566
Bis(tributyltin)oxide [*56-35-9*], *24*:139
2,5-Bis(tributyltin)sulfolane [*41392-14-7*],
    23:136

Bis(3-triethoxysilylpropyl)tetrasulfide [40372-72-3]
  silane coupling agent, 22:150
N,O-Bis(trimethylsilyl)acetamide [10416-59-8]
  silylating agent, 22:144
N,O-Bis(trimethylsilyl)trifluoroacetamide [21149-38-2]
  silylating agent, 22:144
N,N'-Bis(trimethylsilyl)urea [18297-63-7]
  silylating agent, 22:144
Bis(trineophyltin) oxide [60268-17-4], 24:138
Bisulfitation, 23:147
Bitolylene diisocyanate [91-97-4], 24:707
Bitumens
  as roofing material, 21:440
Bituminous sand, 23:719
BJT. See Bipolar junction transistor.
Black chrome, 23:1071
Black glass
  selenium in, 21:712
Black liquor
  from paper mills for fuel, 22:746
Black liquor soap, 23:618
Black masterbatch
  from SBR latex, 22:1005
Black molybdenum, 23:1071
Black pepper
  extraction using SCFs, 23:466
Blackstrap molasses, 23:42
Black tea, 23:753
Blades
  refractory coatings for, 21:106
Blanco Directo process
  for sugar, 23:34
Blanket crepes, 21:565
Blast furnaces
  refractories for, 21:83
Blasting agents
  sodium nitrate for prdn of, 22:392
Blasting caps
  initiator in, 23:113
Blastomycosis, 24:829
Bleaches
  in paper recycling, 21:18
  sampling standards for, 21:627
  sulfamic acid with, 23:130
Bleaching, 23:319
  silicates for, 22:24
Blends
  of natural rubber, 21:578

Bleomycin [11056-06-7], 24:836
Blocking
  in PVAc, 24:970
Bloedite, 22:404
Blood
  analysis of heavy metals in, 22:646
  mannitol as freezing preventative, 23:113
  silver cmpds in analysis of, 22:190
  trace analysis of, 24:508
  use of ultrafiltration, 24:622
Blood glucose, 21:819
  for sugar alcohols, 23:107
  sweetener effect on, 23:557
Blowing agents, 22:52
  PVC as, 24:1036
  rubber chemicals, 21:460
BMC. See Bulk molding compound.
Body washes, 22:322
Boehmite [1318-23-6]
  from sol–gel technology, 22:520
Bohlin rheometer, 21:394
Boilers
  refractories for, 21:83
  sulfamic acid cleaner for, 23:129
Bolometers
  for optical spectroscopy, 22:635
Bolts, explosive
  shape-memory alloy connectors, 21:969
Bombykol, 21:822
Bombyx mori, 22:155; 23:549
Bond-Elut, 24:493
Bond work index
  particle size reduction, 22:281
Bone repair
  silk for, 22:162
Bone screws
  tantalum for, 23:671
Bookbinding
  sorbitol in, 23:112
Borates
  of poly(vinyl alcohol), 24:989
Borax
  centrifugal separation of, 21:867
Borax stability, 24:964
Boric acid
  centrifugal separation of, 21:867
  use in nuclear power steam cycle, 22:754
Borides
  as refractory coatings, 21:106
Borneol [507-70-0], 23:835

Bornyl salicylate [560-88-3], 21:615

Borol process, 23:318

Boron [7440-42-8], 21:119
in brazing filler metals, 22:484
dopant for silicon, 21:1098
role in steelmaking, 22:779
in sodium nitrate, 22:390
thin films of, 23:1060
vitreous silica impurity, 21:1036

Boron carbide
thin films of, 23:1061

Boron nitride [10043-11-5], 21:58, 119
refractory coating, 21:113
as tool material, 24:444

Boron nitride fibers, 21:122

Boron oxide
solubility in steam, 22:728

Boron tribromide
dopant for pure silicon, 21:1095

Borosilicate
as sampling probe, 21:634

Borosilicate glass
attachment to vitreous silica, 21:1041

Botrytis cinerea
inhibited by sorbates, 22:579

Bottles
PVC use, 24:1040

Bouguer-Lambert-Beer law
of absorption, 22:628

Bovine brain prolyl-isomerase crystals,
22:622

Bovine serum albumin
in water-in-CO_2 microemulsions, 23:462

Bowie & Dick test, 22:841

BPCT. See Best practicable control
technology.

Brabender instruments, 21:394

Braids
in reinforced plastics, 21:195

Brake facings
use of selenium in, 21:711

Brakes
titanates in, 24:252

Bralon, 23:543

Brass
corrosion rates in acid, 23:122
dielectric constants of, 22:761
thin films of, 23:1071
use of selenium in, 21:711

Brassinolide [72962-43-7], 22:863

Braunite
magnetic intensity, 21:876

Bravoite [12172-92-8], 24:783

Brazing alloys
silver in, 22:174

Brazing filler metals, 22:482

Brazzein [160047-05-2]
potential sweetener, 23:573

Breakage distribution function
particle size reduction, 22:281

Brettanomyces clausenii
inhibited by sorbates, 22:579

Brettanomyces versatilis
inhibited by sorbates, 22:579

Brevotoxins, 24:505

Brewery gases
sulfur removal from, 23:444

Brick, 21:47; 23:129
sampling standards for, 21:627

Bridgman growth experiments, 22:623

Bright stock, 23:163

Brij
ethoxylated alcohol surfactant, 23:508

Brimstone, 23:232

Brine
refractories from, 21:50
source of salt, 22:360
steel quenching in, 22:801

Brinell hardness
carbon eutectoid steel, 22:796
silver, 22:165
of steel, 22:790
of tempered martensite, 22:798
of thallium, 23:953
of tin, 24:107

Brink impactor
for sampling, 21:637

Briquetting, 22:238

British gums, 22:710, 713

Brix refractometer
in sugar analysis, 23:14

Bromat
surfactant, 23:529

Brominated alkylphenol formaldehyde
resin [112484-41-0]
cross-linking agent, 21:470

Bromine [7726-95-6], 24:866
boiling point of, 23:627
specific gravity, 23:625

Bromochloroethane [25620-54-6]
vinylidene chloride from, 24:884

Bromoform [75-25-2], 24:856

16-Bromohexadecylsilane
self-assembled monolayers of, 23:1092

4-Bromo-3-hydroxybenzoic acid [14348-38-0], 21:619

4-Bromomethyl-7-methoxy-1,2-dihydronaphthalene [83747-47-1], 22:888

2-Bromo-3-methylthiophene [14282-76-9], 24:39

Bromophenol blue, 22:418

Bromopyrogallol red [16574-43-9]
  silver detection, 22:187

N-Bromosuccinimide [128-08-5], 22:1079

2-Bromosulfolane [29325-66-4], 23:135

2-Bromothiophene [1003-09-4], 24:38, 39, 41, 45

3-Bromothiophene [872-31-1], 24:38, 39, 43, 45

Bronchography
  use of tantalum in, 23:671

Bronze
  damping of, 21:963
  silicon, 21:1109
  thin films of, 23:1071
  tin alloys as, 24:117

Brookfield viscometer, 21:394

Brookite [12188-41-9], 24:235

Brown crepes, 21:565

Brown dyes
  stilbene dyes, 22:928

Brown seaweed
  mannitol in, 23:97

Brown sugar, 23:25, 41

Brucite [1317-43-7], 21:55

Bruise energy
  in tires, 24:177

B-stage, 21:197

BTX processing
  use of sulfolane, 23:138

Bucket-wheel excavators, 23:731

Bufalin [465-21-4], 22:866

Bug Barrier
  insect repellent, 21:242

Built-up roofing (BUR), 21:438

Bulk molding compound (BMC), 21:198

Bulk sweeteners, 23:74

Bulk viscosity, 21:364

Bumper pads
  precipitated silica in, 21:1025

Buna process
  sodium catalyst for, 22:345

Buna S, 22:995

Buna technology, 22:994

Bünnagel formula, 23:14

Bunte's salt, 24:65

Buprion [34911-55-2], 22:945

BUR. See Built-up roofing.

Burkeite, 22:404

Burning tires. See Tire-derived fuel.

Bus bars
  brazing filler metals for, 22:490

Butacite, 24:932

Butadiene, 23:139
  SBR from, 22:996

1,2-Butadiene
  sodium catalyst for, 22:345

1,3-Butadiene [106-99-0], 24:853

Butadiene copolymers
  with styrene, 22:1024

1,4-Butanediol [110-63-4], 23:551

Butanesulfonic acid [30734-86-2], 23:194

1,2,3,4-Butane tetracarboxylic acid [1703-58-8], 22:1080

1-Butanethiol [109-79-5], 24:21

2-Butanethiol [513-53-1], 24:21

1-Butanol [71-36-3]
  as solvent, 22:536

2-Butanol [78-92-2]
  as solvent, 22:536

Butter
  extraction of lipids, 23:467

Butvar resins, 24:929

N-Butylacetanilide [91-49-6], 21:240

n-Butyl acetate [123-86-4]
  as solvent, 22:542

sec-Butyl acetate [105-46-4]
  as solvent, 22:542

sec-Butyl alcohol
  prdn using SCF, 23:465

Butylamine [109-73-9]
  as solvent, 22:540

tert-Butylamine [75-64-9]
  as solvent, 22:540

Butylate
  thiols in, 24:29

Butyl-based sealants, 21:661

N-tert-Butyl-2-benzothiazolesulfenamide (TBTS) [95-31-8], 21:465

N-tert-Butyl-2-benzothiazolesulfenimide (TBSI) [3741-80-8], 21:465

n-Butyl-4,4-bis(t-butylperoxy)valerate [995-33-5]
  cross-linking agent, 21:470

4-tert-Butylcatechol (TBC) [98-29-3]
  styrene inhibitor, 22:973

n-Butyl-3,5-diiodo-4-hydroxybenzoate [51-38-7], 24:98

t-Butyldimethylsilylimidazole [*54925-64-3*]
  silylating agent, *22*:144
n-Butyl ether [*142-96-1*]
  as solvent, *22*:548
t-Butylethylbenzene
  extraction of *t*-butylstyrene, *23*:138
Butyllithium [*109-72-8*], *24*:853
t-Butyllithium, *22*:59
n-Butyl 3-mercaptopropionate [*16215-21-7*], *24*:7
t-Butylperbenzoate [*614-45-9*]
  cross-linking agent, *21*:470
n-Butyl propionate [*590-01-2*]
  as solvent, *22*:542
sec-Butyl propionate
  as solvent, *22*:542
Butyl rubber, *21*:487
  selenium in, *21*:713
5-*t*-Butylsalicylic acid [*16094-31-8*], *21*:605
Butylstannoic acid [*2273-43-0*], *24*:147
4-*tert*-Butylstyrene (TBS) [*1746-23-2*], *22*:985
sec-Butyl sulfate [*3004-76-0*], *23*:414
n-Butyl thioglycolate [*10047-28-6*], *24*:7
Butylthiostannoic acid [*26410-42-4*], *24*:145
Butyltin trichloride [*1118-46-3*], *24*:138
Butyltin tris(isooctyl thioglycolate) [*25852-70-4*], *24*:13
t-Butyltrichlorosilane [*18171-74-9*], *22*:59
Butyltricyclohexyltin [*7067-44-9*], *24*:138
n-Butyl vinyl ether [*111-34-2*], *24*:1054
t-Butyl vinyl ether [*926-02-3*], *24*:1056
Butyraldehyde [*123-72-8*]
  PVB resins from, *24*:932
Butyraldehyde–aniline reaction [*34562-31-7*], *21*:468
γ-Butyrolactone [*96-48-0*]
  as solvent, *22*:548
  from succinic anhydride, *22*:1077
n-Butyronitrile [*109-74-0*]
  as solvent, *22*:546
Buxamine-E [*14317-17-0*], *22*:865
Buxaprogenstine [*113762-72-4*], *22*:864

C

CAA. See *Clean Air Act.*
CAAA. See *Clean Air Act Amendments.*
Cabbage
  sulfoxides in, *23*:217
Cable
  PVC use, *24*:1040
Cabling, *24*:166
Cab-O-Sil, *21*:998
Cachou de Laval
  sulfur dye, *23*:342
Caddis fly
  silk source, *22*:156
Cadmium [*7440-43-9*]
  in brazing filler metals, *22*:494
  plating with methanesulfonic acid, *23*:326
  regulatory level, *21*:160
  separation by reverse osmosis, *21*:328
  thin films of, *23*:1073
  vapor pressure of, *23*:1046
Cadmium alloys
  as shape-memory alloys, *21*:964
Cadmium arachidate
  thin films of, *23*:1081
Cadmium diethyldithiocarbamate (CdDEC) [*14239-68-0*], *21*:465
Cadmium oxide, *22*:176
Cadmium sulfoselenide [*11112-63-3*], *21*:698
Cadmium telluride [*1306-25-8*], *21*:767; *22*:473; *23*:791
CAER. See *Community Awareness and Emergency Response.*
Caffeic acid
  in sunflower meal, *22*:598
Caffeine [*58-08-2*], *22*:401, 935; *23*:749
  removal using SCFs, *23*:466
CAIR. See *Comprehensive Assessment Information Rule.*
Calamide
  diethanolamine surfactant, *23*:520
Calaverite [*37043-71-3*]
  tellurium in, *23*:783
Calcimar, *24*:102
Calcitonin, *24*:89
Calcitriol [*3222-06-3*], *22*:857
Calcium
  complexes with sugar alcohols, *23*:103
  diffusion coefficient in vitreous silica, *21*:1047
  role in steelmaking, *22*:779
  in silicon and silica alloys, *21*:1106
  sulfur removal, *22*:781
  in tea, *23*:751
  vitreous silica impurity, *21*:1036
Calcium acrylate [*6292-01-9*]
  in grouts, *22*:455

Calcium alkoxide
  use in silica sol–gel systems, 22:527
Calcium aluminate [12042-78-3], 21:54
Calcium aluminate cement
  refractories from, 21:50
Calcium amalgam, 22:339
Calcium–barium–silicon, 21:1119
Calcium carbide [75-20-7], 24:857
Calcium carbonate [471-34-1]
  cause of boiler tube scale, 22:738
  membrane fouling, 21:315
  recovery of, 21:864
Calcium chloride [10043-52-4]
  salt additive, 22:368
  sampling standards for, 21:627
  in silicate grouts, 22:453
  for silk processing, 22:159
  in sodium chloride prdn, 22:361
  solubility in steam, 22:728
Calcium cyclamate [139-06-0], 23:566
Calcium hydroxide, 22:744
  in sodium chloride prdn, 22:361
Calcium hypochlorite
  centrifugal separation of, 21:867
Calcium ions
  role in sweetness, 23:578
Calcium nitrate [10124-37-5], 21:480
  use in silica sol–gel systems, 22:527
Calcium nitride
  solubility in sodium, 22:330
Calcium oxide [1305-78-8]
  as refractory material, 21:57
  as rubber chemical, 21:473
  solubility in sodium, 22:330
  use in silica sol–gel systems, 22:527
Calcium poly(styrene-sulfonate)
  sugar separation on, 23:98
Calcium precipitates
  in beet sugar processing, 23:54
Calcium ricinoleate [6865-33-4]
  as release agent, 21:210
Calcium salicylate dihydrate [824-35-1],
  21:612
Calcium selenite [13780-18-2], 21:698
Calcium silicate
  salt additive, 22:368
Calcium silicide, 22:44
Calcium–silicon, 21:1119
Calcium soaps
  dispersing agents for, 23:209
Calcium–sodium alloys, 22:347
Calcium stannate [12013-46-6], 24:128

Calcium stearate
  in suture coating, 23:547
Calcium sulfamate, 23:131
Calcium sulfate, 22:365; 23:447
  solubility in steam, 22:728
  in tofu prprn, 22:615
Calcium sulfide [20548-54-3], 23:280
  in steel, 22:816
Calcium sulfite, 23:447
Calcium thioglycolate [814-71-1], 24:8
Calcium titanate [12049-50-2], 24:252
Calcium vanadate [14100-64-2], 24:806
Calculators
  solar energy for, 22:475
Calendering
  PVC use, 24:1040
Calfoam, 23:502
  alcohol surfactant, 23:502
Caliche, 22:383
Calorimeters
  detectors for beam spectroscopy, 22:657
Calsoft
  alkylbenzensulfonate surfactant, 23:496
Camay, 22:323
Cambendazole [26097-80-3], 24:831
Camellia sinensis, 23:746
Campesterol [474-62-4]
  in soybeans and other oilseeds, 22:597
Camphene [79-92-5], 23:834
(−)-Campholenic aldehyde [4501-58-0],
  23:861
Camphor [76-22-2], 23:835
Cancer
  vaccines against, 24:743
Cancer chemotherapy
  for animals, 24:836
Candida albicans
  inhibited by sorbates, 22:579
Candida krusei
  inhibited by sorbates, 22:579
Candida mycoderma
  inhibited by sorbates, 22:579
Candida tropicalis
  inhibited by sorbates, 22:579
Candy
  corn syrups in, 23:600
  sugar alcohols in, 23:108
Cane sugar, 23:20
Canfieldite [12250-27-0], 24:105, 123
Cannizzaro reaction, 23:97
Cannon-Fenske viscometer, 21:376
Cansolv process, 23:449

Capacitors
    refractory coatings for, *21*:113
    silver in, *22*:176
    strontium compounds as, *22*:955
    tantalum powders for, *23*:667
    titanium dioxide for, *24*:238
Capillary electrophoresis, *24*:506
    in trace analysis, *24*:492
Capillary tubing
    of vitreous silica, *21*:1064
Capillary viscometers, *21*:375
Capric acid [*334-48-5*]
    separation coefficients for, *21*:926
Caproic acid [*142-62-1*]
    separation coefficients for, *21*:926
Caprolactam
    use of sulfuric acid, *23*:397
N,N'-Caprolactam disulfide [*23847-08-7*]
    cross-linking agent, *21*:470
Caprolactone [*502-44-3*], *23*:547
Caprylic acid [*124-07-2*]
    separation coefficients for, *21*:926
Captafol [*2425-06-1*], *23*:275, 291
Captan [*133-06-2*], *23*:274
Caramel
    molasses in, *23*:604
    vanillin in, *24*:820
Carbadox [*6804-07-5*]
    veterinary drugs, *24*:827
Carbamates
    degradation in soil, *22*:430
    from poly(vinyl alcohol), *24*:990
    textile finishing agent, *23*:904
    as veterinary drugs, *24*:830
Carbide fibers, *21*:117
Carbides
    for centrifuges, *21*:847
    as refractory coatings, *21*:102
    Rockwell hardness of, *24*:397
    in steel, *22*:794
Carbimazole [*22232-54-8*], *24*:99
Carbohydrates
    in soybeans and other oilseeds, *22*:598
    sulfation of, *23*:171
Carbohydrazide
    oxygen scavenger for steam, *22*:742
Carbolic acid
    specific gravity, *23*:625
Carbometalation, *24*:310
Carbon [*7440-44-0*], *21*:56, 119; *23*:1046;
        24:32
    emissivities of, *23*:827

    in ferrous shape-memory alloys, *21*:965
    as refractory material, *21*:57
    in steel, *22*:765, 775
    vapor pressure of, *23*:1046
Carbonated beverages
    high fructose syrups in, *23*:597
    stimulants in, *22*:936
    sweeteners for, *23*:558
Carbon bisulfide
    boiling point of, *23*:627
Carbon black
    PVC pigment, *24*:1035
    sampling standards for, *21*:627
Carbon dioxide [*124-38-9*], *24*:853
    for decaffeination of tea, *23*:762
    as refrigerant, *21*:131, 134
    removal of, *23*:436
    supercritical fluid, *23*:454
Carbon diselenide [*506-80-9*], *21*:699
Carbon disulfide [*75-15-0*], *23*:267; *24*:831
    removal of, *23*:436
    specific gravity, *23*:625
    from sulfur, *23*:237, 259
Carbonless copypaper
    using salicylic acid, *21*:611
Carbon monosulfide [*2944-05-0*], *23*:267
Carbon monoxide [*630-08-0*], *24*:852
    effect on refractories, *21*:81
    refractories for, *21*:79
    sensors for, *21*:823
Carbon steel, *22*:808
    brazing filler metals for, *22*:490
    for oleum, *23*:391
    for tanks and pressure vessels, *23*:643
    as tool materials, *24*:392
Carbon subsulfide [*12976-57-2*], *23*:267
Carbon sulfotelluride [*10340-06-4*], *23*:796
Carbon tetrachloride [*56-23-5*], *24*:832,
        865
    as HAP compound, *22*:532
    regulatory level, *21*:160
    role in sol–gel technology, *22*:515
    as solvent, *22*:538
Carbonyl sulfide [*463-58-1*], *23*:267, 270,
        301
    removal of, *23*:436
Carbonyl telluride [*65312-92-7*], *23*:796
Carbowax, *23*:673
3-(4-Carboxybenzoyl)-2-quinoline-
        carboxaldehyde, *24*:517
Carboxylates, *23*:492
Carboxylate soaps, *22*:299

Carboxylic acids
    reactions with silanes, 22:51
Carboxymethylcellulose
    in de-inking, 21:14
    water additive of steam prdn, 22:740
Carboxymethylthiosuccinic acid [99-68-3],
        24:2
Carburizing
    of steel, 22:808
Carcinogens
    analysis using fluorescence, 22:653
Cardiac steroids, 22:852
CarDio process, 24:711
Cardiovascular agents
    steroids as, 22:852
    thiophene and thiophene derivatives
        as, 24:47
(+)-2-Carene [4497-92-1], 23:843
3-Carene [13466-78-9], 23:834
Caress, 22:323
Carmack Amendment, 24:537
Carman-Kozeny equation, 21:670
Carnauba wax [8015-86-9]
    as release agent, 21:210
    release substrate, 21:212
Carnot cycles, 22:749
Carnot efficiency
    role in thermoelectric energy
        conversion, 23:1032
Carnotite [1318-26-9], 24:783
Carnot's equations, 23:988
Caroflex MP, 24:1064
Caro's acid, 23:299, 312
β-Carotene [7235-40-7], 23:874
    from SCF solutions, 23:472
γ-Carotene [472-93-5], 23:874
Carotenoids, 24:504
    as terpenes, 23:874
Cars. See Coherent anti-Stokes Raman
        spectroscopy.
Carsonol
    α-olefinsulfonate surfactant, 23:497
Carsonol MLS
    surfactant, 23:501
Carsonol TLS 65
    surfactant, 23:501
Carsul, 23:267
Cartilage wire
    tantalum for, 23:671
Carvone [99-49-0], 23:835
(−)-Carvone [6485-40-1], 23:845
β-Caryophyllene [87-44-5], 23:836

Cascade cycles
    in refrigeration, 21:139
Case hardening
    of steels, 21:103; 22:808
Cashmere Bouquet, 22:323
Cassava, 24:98
    starch from, 22:699
Cassiterite [1317-45-9], 24:105, 122
Casson-Asbeck plots, 21:354
Casson equation, 21:350
Casson plots, 21:351
Casting
    of steel, 22:783
Cast iron
    damping of, 21:963
    dielectric constants of, 22:761
    emissivities of, 23:827
    selenium in prdn of, 21:710
    silicon in, 21:1113
    strontium in, 22:950
    tellurium in prdn of, 23:801
    tin alloys in, 24:120
    tool materials for, 24:405, 432
Castner cell
    for sodium, 22:335
Catacarb, 23:439
Catalysis
    analysis by Mössbauer spectroscopy,
        22:656
Catalysts
    centrifugal separation of, 21:861
    for grouting systems, 22:454
    heterogeneous tin, 24:127
    petroleum reforming, 21:342
    silicates as, 22:25
    silver cmpds, 22:168, 190
    tellurium in, 23:804
    thioglycolic acid as, 24:12
    titanates as, 24:290
    vanadium compounds as, 24:809
Catalyst Stabilization Technology (CST),
        22:972
Catalyst supports
    silica gels as, 21:1023
    silica sols as, 21:1000
    using silylating agents, 22:151
Catalytic converters
    brazing filler metals for, 22:490
Catechins
    in tea, 23:747
Catechol, 24:79
Catecholamines, 24:507

Catechol sulfate [4074-55-9], 23:414
Cathode ray tubes
    strontium carbonate in, 22:951
Cathodic protection
    solar energy for, 22:475
    for tanks, 23:655
    thermoelectric power supplies for, 23:1035
Cationic starches, 22:710, 714
Cationic surfactant, 23:523
Caulking compounds
    talc in, 23:614
Caustic soda
    tanks for, 23:647
Cavity ringdown laser absorption
    spectroscopy, 22:659
CBER. See Center for Biologics
    Evaluation and Research.
CBI. See Confidential business
    information.
CBTS. See N-Cyclohexyl-2-
    benzothiazolesulfenamide.
CdDEC. See Cadmium
    diethyldithiocarbamate.
CDER. See Center for Drug Evaluation
    and Research.
CDRH. See Center for Devices and
    Radiological Health.
CED. See Cohesive energy density.
α-Cedrene [469-61-4], 23:869
Cedrol [77-53-2], 23:869
Cedryl acetate [61789-42-2], 23:869
Cedryl methyl ether [19870-74-7], 23:869
Cefluofor
    as veterinary drugs, 24:827
Cefoxitin [35607-66-0], 24:48
Celery
    mannitol in, 23:97
Celestite [14291-02-2], 22:949, 950
Celite, 22:191
    silylating agents for, 22:148
Cell growth
    silk for, 22:162
Cell-mediated immunity, 24:740
Cellophane [9005-81-6]
    as release agent, 21:210
    release films, 21:214
    sodium sulfide in prdn of, 22:417
    sulfur in prdn of, 23:259
Cellophane, coated
    for tea packaging, 23:760
Cells
    of vitreous silica, 21:1065

Cellulose
    dextrose from, 23:583
    DMSO as solvent for, 23:228
    sulfated using sulfamic acid, 23:125
Cellulose acetate
    membrane for ultrafiltration, 24:604
    in reverse osmosis membranes, 21:304
    sulfur in prdn of, 23:259
Cellulose hollow fibers
    sulfolane as plasticizer, 23:139
Cellulose triacetate
    in reverse osmosis membranes, 21:304
Cellulose xanthate
    in soil conditioning, 22:458
Cement
    sampling standards for, 21:627
    silicates in prdn of, 22:23
Cementation
    of refractory coatings, 21:102
Cemented carbides, 24:404
    brazing filler metals for, 22:490
    tantalum carbide in, 23:676
Cementite, 22:789
Cement kilns
    refractories for, 21:83
    scrap tire fuel, 21:24
Center for Biologics Evaluation and
    Research (CBER), 21:170
Center for Devices and Radiological
    Health (CDRH), 21:170
Center for Drug Evaluation and Research
    (CDER), 21:170
Center for Food Safety and Applied
    Nutrition (CFSAN), 21:171
Center for Veterinary Medicine (CVM),
    21:171
Centrifugal atomizer, 22:672
Centrifugal casting
    reinforced plastics, 21:203
Centrifugal separation, 21:828
Centrifugal sucrose, 23:75
Centrifuges
    for separation, 21:829
Cephalosporin P_1 [13258-72-5], 22:869
Cephalosporins
    silylation in synthesis of, 22:146
    as veterinary drugs, 24:827
Cephalosporium sp.
    inhibited by sorbates, 22:579
Cephalostatin 1 [11288-65-9], 22:866
Ceramers
    from sol–gel technology, 22:525

Ceramic films
  in metal oxide sensors, *21*:825
Ceramic glazes
  strontium carbonate in, *22*:953
Ceramics, *24*:127
  bioactive sol–gel materials, *22*:526
  for centrifuges, *21*:847
  for cooling tower packing, *22*:213
  as refractory coatings, *21*:101
  Rockwell hardness of, *24*:397
  SCF solution of, *23*:472
  selenium as colorant for, *21*:712
  silica for, *21*:999
  talc application, *23*:613
  tellurium as colorants for, *23*:804
  thermocouple insulation, *23*:823
  thorium in manfg of, *24*:71
  as tool materials, *24*:427
  triarylmethane dyes for, *24*:565
Ceramic tile
  sulfur impregnated, *23*:263
Cerargyrite [*14358-96-4*], *22*:170
Cerasynt
  glycerol ester surfactant, *23*:512
CERCLA. See *Comprehensive
    Environmental Response,
    Compensation, and Liability Act.*
Cercospora sp.
  inhibited by sorbates, *22*:579
Cereals
  dextrose for, *23*:593
  sucrose in, *23*:6
Cerfak
  ethoxylated alcohol surfactant, *23*:508
Ceric ammonium nitrate
  silver detection, *22*:187
Ceric sulfate
  for chemical dosimeters, *22*:845
Cerium
  in magnesium ferrosilicon, *21*:1117
  thorium oxides, *24*:76
Cerium tellurium molybdate, *23*:804
Cer Max 460, *24*:430
Cermets
  as refractory coating, *21*:91
  tantalum nitride in, *23*:676
  as tool materials, *24*:414
Cesium
  diffusion coefficient in vitreous silica,
    21:1047
  rubidium as by-product, *21*:593

Cesium chloride
  use in density gradient separation,
    21:853
Cesium hexachlorotitanate [*16918-47-1*],
    24:261
Cestodes
  veterinary drugs against, *24*:832
Cetyltrimethylammonium bromide
  use in silica analyses, *21*:1012
CFC. See *Chlorinated fluorocarbon.*
CFSAN. See *Center for Food Safety and
    Applied Nutrition.*
cGMP. See *Current Good Manufacturing
    Practice.*
Chaetomium globosum
  inhibited by sorbates, *22*:579
Chain-transfer agents
  thioglycolic acid as, *24*:12
Chain-transfer constants
  in free-radical styrene polymerization,
    22:1040
Chalcedony, *21*:1028
Chalcone isomerase, *23*:751
Chalcone synthase, *23*:751
Chalcopyrite [*1308-56-1*]
  sulfide ore, *23*:244
  tellurium in, *23*:782
Chalk-Harrod mechanism
  for hydrosilylation, *22*:96
Chamomile
  extraction using SCFs, *23*:466
Charcoal
  for silicon prdn, *21*:1105
Charcoal briquettes
  sodium nitrate for prdn of, *22*:393
Cheese slices
  sorbates in, *22*:582
Chelating agents
  for reverse osmosis pretreatments,
    21:315
  role in SCF usage, *23*:468
  in soap, *22*:320
Chelators
  sucrose carboxylates as, *23*:73
Chemical grouting, *22*:452
Chemical Manufacturers Association,
    21:154
Chemical oxygen demand, *21*:165
Chemical potential
  in thermodynamics, *23*:996
Chemical process industry
  regulatory agencies, *21*:154

Chemicals
  transportation of, *24*:524
Chemical treatments
  in textile finishing, *23*:891
Chemical vapor deposition, *21*:771
  for diamond films, *24*:439
  of thin carbide films, *24*:230
  for thin films, *23*:1060
  for tool coatings, *24*:418
Chemiluminescence, *22*:654
Chemometrics
  use in optical spectroscopy, *22*:641
Chemotherapeutics, anticancer
  as veterinary drugs, *24*:836
Chem-Trete, *22*:79
Chenodeoxycholic acid [*474-25-9*], *22*:856
Cherries, *23*:96
Chert, *21*:992, 1028
Chewing gum
  flavor enhancer for, *23*:573
  SBR use, *22*:1011
  sugar alcohols in, *23*:108
Chicle
  purification by centrifuge, *21*:857
Chief Technology Officer, *21*:264
Chilean nitrate, *22*:383
Chilean saltpeter, *22*:383
Chilies
  extraction using SCFs, *23*:466
Chilled surface drying
  of soap, *22*:315
Chillproofing, *21*:1001
China clay
  thickening of, *21*:680
Chinaware
  selenium as colorant for, *21*:712
Chirality
  analysis using optical spectroscopy,
    22:651
  in titanium complexes, *24*:300
Chiral separation, *24*:500
Chitosan
  dyes from sulfonic acid complexes,
    23:207
Chiyoda process, *23*:739
Chlamydia sp.
  veterinary drugs against, *24*:829
Chloral [*75-87-6*], *24*:865
Chlor–alkali production
  sodium chloride in, *22*:374
Chlorargyrite, *22*:179
Chlordane [*12789-03-6*], *22*:425

4,7-Chlordane [*59-74-9*], *22*:421
  regulatory level, *21*:160
Chloride
  silver cmpds in detm of, *22*:190
  in steam, *22*:737
Chloride process
  for titanium dioxide, *24*:241
Chlorimuron, *24*:510
Chlorimuron ethyl [*90982-32-4*], *22*:448
Chlorinated fluorocarbon (CFC), *21*:139,
    165
Chlorinated polyethylene, *21*:489
Chlorinated PVC, *24*:1039
Chlorination
  selenium recovery by, *21*:694
  for tantalum separation, *23*:663
  of toluene, *24*:357
Chlorine [*7782-50-5*], *21*:695; *24*:852
  diffusion in vitreous silica, *21*:1048
  effect on reverse osmosis membranes,
    21:315
  to make PVC, *24*:1043
  refractories for, *21*:80
  role in sol–gel technology, *22*:515
Chlorine atom [*22537-15-1*], *24*:854
Chlorine dioxide
  in paper recycling, *21*:18
Chloroacetaldehyde [*107-20-0*], *24*:855
Chloroacetic acid [*79-11-8*], *24*:2
2-Chloroacrylonitrile [*920-37-6*], *24*:3
Chloroaniline
  separation by reverse osmosis, *21*:319,
    323
Chlorobenzene [*108-90-7*]
  as HAP compound, *22*:532
  regulatory level, *21*:160
  as solvent, *22*:538
Chlorocarbonylsulfenyl chloride [*2757-
    23-5*], *23*:273
Chlorodifluoromethane
  as SCF, *23*:472
2-Chloroethanol [*107-07-3*], *24*:865
1-Chloroethylbenzene [*672-65-1*], *24*:856
2-Chloroethyl hydrogen sulfate [*36168-
    93-1*], *23*:417
2-Chloroethyl vinyl ether [*110-75-8*],
    24:1056
Chlorofluorocarbons
  as refrigerant, *21*:131, 134
1-Chloro-1-fluoro-2-nitroethane [*461-70-1*],
    24:856
Chloroform [*67-66-3*], *24*:856
  boiling point of, *23*:627

dielectric constants of, 22:759
as HAP compound, 22:532
potential sweetener, 23:573
regulatory level, 21:160
as solvent, 22:538
Chlorogenic acid
in sunflower meal, 22:598
Chloromethylation
of toluene, 24:357
Chloromethyldimethylchlorosilane [1719-57-9]
silylating agent, 22:144
5-Chloromethylsalicylic acid [10192-87-7], 21:606
2-Chloro-3-methylthiophene [14345-97-2], 24:48
2-Chloromethylthiophene [765-50-4], 24:40
Chlorophenols, 23:298
rejection by reverse osmosis membrane, 21:319
N-(2-Chlorophenyl)-1-chloromethane sulfonamide [30064-44-9], 23:209
p-Chlorophenyl vinyl ether [1074-56-2], 24:1056
Chlorophyll [479-61-8], 23:258, 873
Chloroprene [126-99-8], 24:856
2-Chloropropane [75-29-6], 24:855
2-Chloro-1-propanol [78-89-7], 24:856
2-Chloropropionaldehyde [683-50-1], 24:856
2-Chloropropionic acid [598-78-7], 24:6
3-Chloropropylmethylchlorosilane [33687-63-7], 22:38
3-Chloropropyltrimethoxysilane [25512-39-4]
critical surface tension of, 22:148
silane coupling agent, 22:150
Chlorosilanes, 22:39; 23:139
Chlorostannites, 24:124
N-Chlorosuccinimide [128-09-6], 22:1079
N-Chlorosulfamic acid [17172-27-9], 23:123
3-Chlorosulfolane [3844-04-0], 23:135
Chlorosulfonated polyethylene (CSPE) [9008-08-6], 21:489; 23:298
as roofing material, 21:446
Chlorosulfonation, 23:302
of paraffin sulfonates, 23:172
Chlorosulfuric acid [7790-94-5], 23:195, 203, 289, 297
for sulfation, 23:147
Chlorothalonil [1897-45-6], 22:422

p-Chlorotoluene [106-43-4]
as solvent, 22:538
6-Chloro-D-tryptophan [17808-35-4]
potential sweetener, 23:573
1-Chlorovinyl radical [50663-45-1], 24:856
2-Chlorovinyl radical [57095-76-8], 24:856
Chlorpyrifos [2921-88-2], 22:422
Chlorsulfuron [64902-72-3], 22:448; 24:510
Chlortetracycline [57-62-5], 24:829, 831
Chocolate
vanillin in, 24:819
Cholane [548-98-1], 22:854
5η-Cholanic acid [546-18-9], 22:855
Cholera
vaccine against, 24:742
Cholestane [145-53-7], 22:854
Cholesterol [57-88-5], 22:852
removal using SCFs, 23:466
solubility in carbon dioxide, 23:457
Cholic acid [81-25-4], 22:852
Choline salicylate [2016-36-6], 21:613
Chromatographic methods
for sugar analyses, 23:17
Chromatography, 21:310; 24:743
of black tea, 23:756
of sugar alcohols, 23:103
use of silica gels, 21:1001, 1023
use of silylating agents, 22:143, 149
Chromatography columns
coatings for, 22:79
vitreous silica, 21:1065
Chrome cake, 22:403
Chromel-alumel
thermocouples, 23:823
Chromel-constantan
thermocouples, 23:823
Chrome–nickel alloys, 22:770
Chrome ore
refractories from, 21:50
Chrome plating
for centrifuges, 21:847
Chrome–tin pink, 24:127
Chrome titanate buff, 24:248
Chromia [1308-38-9], 24:854
Chromides
as refractory coatings, 21:103
Chromite [53293-42-8], 21:83
magnetic intensity, 21:876
Chromium [7440-47-3], 23:1046
in brazing filler metals, 22:492
brazing of, 22:484
emissivities of, 23:827

in ferrous shape-memory alloys, *21*:965
regulatory level, *21*:160
role in steelmaking, *22*:779
separation by reverse osmosis, *21*:322
steel for steam turbines, *22*:749
in steels, *22*:815
in tool steels, *24*:398
vapor pressure of, *23*:1046
vitreous silica impurity, *21*:1036
Chromium–iron alloys
  steels from, *22*:817
Chromium–molybdenum, *22*:748
Chromium plating
  use of selenium in, *21*:711
Chromium salts
  for suture treatment, *23*:545
Chromizing
  of iron, *21*:103
Chromophores
  role in optical spectroscopy, *22*:644
Chrysene
  analysis using laser-induced
    fluorescence, *22*:654
Chrysocolla
  magnetic intensity, *21*:876
Chrysotile
  sampling standards for, *21*:627
Chuca, *22*:383
CI 18790, *23*:343
CI 24890, *22*:924, 925
CI 24895, *22*:926, 929
CI 40000, *22*:922, 926
CI 40003, *22*:926
CI 40006, *22*:926
CI 40030, *22*:923
CI 40065, *22*:923
CI 40215, *22*:923
CI 40505, *22*:923
CI 53000, *23*:342
CI 53005, *23*:351
CI 53010, *23*:351
CI 53040, *23*:350
CI 53045, *23*:350
CI 53050, *23*:342, 344, 350
CI 53065, *23*:350
CI 53090, *23*:350
CI 53160, *23*:341
CI 53165, *23*:351
CI 53166, *23*:351
CI 53180, *23*:342
CI 53185, *23*:342, 343, 345, 352
CI 53186, *23*:343

CI 53210, *23*:351
CI 53228, *23*:348, 353
CI 53230, *23*:342, 352
CI 53275, *23*:351
CI 53280, *23*:351
CI 53290, *23*:342, 353
CI 53320, *23*:342, 350
CI 53325, *23*:342
CI 53327, *23*:342
CI 53335, *23*:351
CI 53430, *23*:342, 352
CI 53440, *23*:352
CI 53570, *23*:342, 352
CI 53571, *23*:352
CI 53630, *23*:342
CI 53640, *23*:342, 353
CI 53710, *23*:342, 352
CI 53720, *23*:353
CI 53722, *23*:351
CI 53800, *23*:342, 353
CI 53810, *23*:342, 353
CI 58820, *23*:354
CI 58825, *23*:354
CI 69700, *23*:354
CI 69705, *23*:354
CI 70305, *23*:354
CI 70310, *23*:354
Cider
  purification by centrifuge, *21*:857
1,4-Cineole [*470-67-7*], *23*:834
1,8-Cineole [*470-82-6*], *23*:834
Cinnamate 4-hydroxylase, *23*:751
Circuit boards
  silver in, *22*:176
  solders for, *22*:487
Circular dichroism, *22*:651
CI Sulfur Brown 1, *23*:341
Citral dimethyl acetal [*7549-37-3*], *23*:863
Citric acid, *21*:317
Citronella [*106-23-0*], *23*:835, 858
Citronellal [*39785-81-4*]
  as mosquito repellent, *21*:244
(+)-Citronellal [*2385-77-5*], *23*:859
Citronellol [*106-22-9*], *23*:835
Citronellyl acetate [*150-84-5*], *23*:858
Citrus oils
  extraction using SCFs, *23*:466
Citrus pulp
  use of sorbates, *22*:584
Cladosporium cladosporiodes
  inhibited by sorbates, *22*:579
Clapeyron equation, *23*:1006

Clarification, 21:667
  by centrifugal separation, 21:861
Claus off-gas treatment (SCOT) process, 23:444
Claus process, 23:278, 301
  for sulfur recovery, 23:440
Clays, 23:727
  centrifugal separation of, 21:850
  nanomaterials from sol–gel technology, 22:526
  particle size measurement for, 22:270
  purification by magnetic separation, 21:891
  refractories from, 21:49
  as release agent, 21:210
  sorbents for pesticides, 22:437
  ultrafiltration of, 24:615
Clean Air Act (CAA), 21:165, 1120
  effect on sulfur, 23:253
Clean Air Act Amendments (CAAA), 21:165; 22:531; 23:309
Cleaners
  precipitated silica in, 21:1025
  sulfamic acid as, 23:129
Cleaning
  in paper recycling, 21:17
  use of SCFs, 23:465
Cleaning solvents, 22:569
Cleanroom technology, 22:276
Clean Water Act (CWA), 21:165
CLEAR process, 23:245
Clerget double polarization
  in sugar analysis, 23:14
Clerici's solution [61971-47-9], 23:956
Climet counter, 22:276
Clinical trials, 21:172
Clintox process, 23:449
Clioxanide [144327-41-3], 24:832
Clomipramine [303-49-1], 22:942
Cloning, 24:741
Clopidogrel [90055-48-4], 24:47
Clostridia
  antimicrobial agent for, 24:828
Clostridium botulinum
  destruction of in foods, 22:848
Clostridium perfringens
  destruction of in foods, 22:848
  inhibited by sorbates, 22:580
Clostridium sporogenes
  inhibited by sorbates, 22:580
Clostridium stricklandii
  selenium in, 21:702

Cloud seeding
  sodium iodide for, 22:191, 382
CMOS circuits, 21:742
Coagulants
  for sedimentation, 21:671
Coagulation, 21:671
  water treatment for steam prdn, 22:745
Coal, 22:250
  centrifugal separation of, 21:847
  magnetic separation of, 21:886
  sampling standards for, 21:627
  for silicon prdn, 21:1105
  for steam prdn, 22:758
  thickening of, 21:680
  transportation of, 24:524
  use of sulfur for, 23:252
Coal gasification
  use of steam, 22:758
Coal liquefaction
  use of steam, 22:758
Coal-tar pitch
  as roofing material, 21:442
Coast, 22:323
Coated felts
  as roofing material, 21:439
Coated Vicryl, 23:547
Coatings
  for baking purposes, 23:559
  clear antifog, 23:208
  corrosion resistant, 24:323
  crystalline diamond, 24:439
  optical interference, 23:1049
  overlay, 21:113
  for precipitated silica, 21:998
  protective, 23:112
  PVB in, 24:935
  PVC use, 24:1040
  refractory, 21:88
  rhenium, 21:341
  as roofing material, 21:440
  silica gels in, 21:1000, 1023
  silicates for, 22:23
  silicon esters in prprn of, 22:79
  silver, 22:164, 176
  sol–gel, 22:499
  solvents for, 22:568
  for steel tool materials, 24:403
  sulfur as, 23:262
  talc for, 23:614
  tin alloy, 24:116
  from titanium alkoxides, 24:285
  for titanium dioxide pigment, 24:247

for tool materials, 24:417
vinylidene chloride copolymer, 24:915
vitreous silica, 21:1069
wear-resistant titanium carbide, 24:231
Coatings, antireflective
thin films for, 23:1073
Cobalt
adhesion in tire cord, 24:171
in brazing filler metals, 22:494
complexes with sugar alcohols, 23:103
in ferrous shape-memory alloys, 21:965
in tool steels, 24:399
vitreous silica impurity, 21:1036
Cobalt alloys
as brazing filler metals, 22:494
Cobalt alloys, cast
Rockwell hardness of, 24:397
as tool materials, 24:404
Cobalt–chrome–tungsten
for centrifuges, 21:848
Cobalt chromite [12016-69-2], 24:855
Cobalt–iron–titanium alloy
use of selenium in, 21:711
Cobalt molybdate
for hydrogen sulfide prdn, 23:281
Cobalt(II) oxide
solubility in steam, 22:728
Cobalt–silicon alloys, 22:58
Cobalt titanate green, 24:248
Cobbers, 21:887
Cocamidopropylbetaine [61789-40-0],
22:322
Cocatalyst
sodium bromide for, 22:379
Coccidiosis, 24:831
Coccidiostat, 24:146
Cocoa bean
mannitol in, 23:97
Coconut
extraction using SCFs, 23:467
sorbitol in, 23:110
Coconut oil [8001-31-8], 22:302
Code of Federal Regulations, 21:165;
23:776
Coe and Clevenger method
for sedimentation flux, 21:673
Coefficient of performance, 23:988
Coenzyme A [85-61-0], 24:20
selenium in, 21:702
Coesite [13778-38-6], 21:981
COFCAW process, 23:730
Coffee
decaffeination process, 23:467

decaffeination using SCFs, 23:465
stimulants in, 22:936
Coffinite [14485-40-6], 24:643
Cogeneration
use of steam, 22:754
Coherent anti-Stokes Raman spectroscopy
(cars), 22:650
Cohesive energy density (CED), 22:534;
23:454
Coins
silver alloys for, 22:175
Coke
role in steelmaking, 22:767
Coke ovens
refractories for, 21:83
Cola
stimulants in, 22:936
Cold flow
in PVAc, 24:970
Cold GR-S, 22:1004
Cold press molding, 21:197
Cold solvent cleaning, 22:569
Collagen
sutures of, 23:545
Colletotrichum lagenarium
inhibited by sorbates, 22:579
Colloids
as foulants for reverse osmosis, 21:313
Colorants
selenium for glass and ceramics,
21:712
in soap, 22:319
Color-center production
is vitreous silica, 21:1062
Color filters, 24:566
Colorimetric methods
in sugar analysis, 23:16
Color matching
in stains, 22:693
Color photography
silver cmpds for, 22:192
Columbite [1306-08-7]
magnetic intensity, 21:876
tantalum in, 23:660
Columbite–tantalite
magnetic intensity, 21:876
Columns, chromatographic. See
Chromatography columns.
Combined cycles, 22:754
Combustion
analysis using laser-induced
fluorescence, 22:653

Combustion aids, 23:212
Commercial starches, 22:699
Communications equipment
  steel for, 22:826
Communication systems
  tantalates in, 23:677
Community Awareness and Emergency
  Response (CAER), 21:165
Complexes
  of PVP, 24:1088
Composite coatings
  for centrifuges, 21:848
Composites, 21:194
  coupling agents for, 24:329
  high performance from sol–gel
    technology, 22:524
  silylating agents for, 22:152
  textile use, 23:882
Composites, lubricating
  tellurium in, 23:804
Compound cycles
  in refrigeration, 21:139
Compound semiconductors, 21:763
Comprehensive Assessment Information
  Rule (CAIR), 21:165
Comprehensive Environmental Response,
  Compensation, and Liability Act
  (CERCLA), 21:165
Compressibility factor, 23:1011
Compression molding, 21:198
Compression tests, 22:227
Compressor
  for refrigeration, 21:129
Computer-assisted tomography
  sodium iodide detectors for, 22:382
Computer chips
  vitreous silica in prdn of, 21:1068
Computers
  synthetic quartz crystals for, 21:1082
Computers, personal
  solid tantalum capacitors in, 23:670
Computer technology
  role in sensor development, 21:817
  use in technical service, 23:779
Concentration
  of tea leaf, 23:761
Conco
  ethoxylated alkylphenol surfactant,
    23:511
Conco Sulfate A
  surfactant, 23:501
Conco Sulfate EP
  surfactant, 23:501

Conco Sulfate P
  surfactant, 23:501
Conco Sulfate WR Dry
  surfactant, 23:501
Concrete
  lignin sulfonate additives for, 23:205
  poly(vinyl acetate) in, 24:974
  prepared with sulfur, 23:262
  sampling standards for, 21:627
  sulfonates in, 23:158
  for tanks and pressure vessels, 23:644
  tool materials for, 24:436
Concrete additives
  lignosulfonates for, 23:169
Condensation
  role in sol–gel technology, 22:505
  simultaneous heat and mass transfer,
    22:198
Condensed whey
  use of sorbates, 22:584
Condensers
  for refrigeration, 21:129
  sulfamic acid cleaner for, 23:129
Condense S Blue 2 [12224-49-6], 23:343
Condense S Orange 2, 23:343
Conduit
  PVC use, 24:1040
Cone angle
  in atomizers, 22:680
Cone crushers, 22:285
Cone–plate viscometer, 21:390
Confectionery products
  dextrose for, 23:593
  high fructose syrup in, 23:597
  polyols in, 23:106
  soybeans and other oilseeds in, 22:610
  sucrose in, 23:6
  sugar alcohols in, 23:108
  vanillin in, 24:820
Confidential business information (CBI),
  21:165
η-Conglycinin
  in soybeans, 22:595
Congo, 21:298
Consolidation by powder metallurgy,
  24:401
Consumer Product Safety Commission
  (CPSC), 21:165
Consumer protection, 21:168
Contact angle, 23:484
Contact lenses
  PVP hydrogels, 24:1078

Continuous laminating
   reinforced plastics, *21*:203
Contraceptive drugs
   steroids as, *22*:852
Controlled-release technology
   role of SCFs, *23*:472
Controlled-stress viscometers, *21*:392
Control panels
   silver switches for, *22*:176
Control Technique Guideline documents
      (CTGs), *22*:531
Conveyor belts
   textile use, *23*:882
Conveyors
   sampling from, *21*:647
Coolanol, *22*:79
Cooling water
   sodium bromide treatment, *22*:379
Coordination complexes
   of thorium, *24*:73
Copaiba [*8001-61-4*], *21*:298
Copal, *21*:298
Copal resin [*9000-14-0*], *21*:297
Copolymer alloy membranes
   as roofing material, *21*:448
Copolymerization
   of VDC, *24*:889
Copolymers
   fractionation using SCFs, *23*:469
   of poly(vinyl acetals), *24*:928
Copper
   adhesion in tire cord, *24*:171
   complexes with sugar alcohols, *23*:103
   corrosion rates in acid, *23*:122
   critical surface tension of, *22*:148
   emissivities of, *23*:827
   plating with methanesulfonic acid,
      23:326
   rhenium as by-product, *21*:336
   role in steelmaking, *22*:780
   as sampling probe, *21*:634
   in scrap for steel, *22*:779
   silver as by-product, *22*:171
   single dc-diode sputtering of, *23*:1053
   sodium sulfides for prdn, *22*:414
   in steam, *22*:737
   sulfuric acid as by-product, *23*:381
   sulfur in ore leaching, *23*:251
   tellurium as by-product, *23*:786
   tellurium in ores, *23*:782
   thickening of, *21*:680
   tool materials for, *24*:436
   vitreous silica impurity, *21*:1036

Copper alloys
   brazing filler metals, *22*:492
   dielectric constants of, *22*:761
   in presence of steam, *22*:743
   silver in, *22*:174
Copper anode
   selenium as by-product of, *21*:690
Copperas, *24*:244
   by-product of titanium dioxide, *24*:246
Copper-constantan
   thermocouples, *23*:823
Copper cyanide
   separation by reverse osmosis, *21*:322
Copper *O,O*-diisopropylphosphorodithioate
      [*41593-12-8*], *21*:468
Copper dimethyldithiocarbamate
      [*137-29-1*], *21*:466
Copper hydroxide [*20427-59-2*], *22*:422
Copper indium diselenide [*12018-95-0*],
      21:715; *22*:473
Copper leaching
   use of sulfuric acid, *23*:397
Copper–lead–tellurium alloys, *23*:803
Copper oxide
   for sulfur dioxide recovery, *23*:446
Copper(II) oxide
   solubility in steam, *22*:728
Copper oxides
   cause of boiler tube scale, *22*:738
   solubility in steam, *22*:729
Copper phthalocyanine
   grown in space, *22*:624
Copper–selenium alloys, *21*:710
Copper–silicon alloys, *22*:58
Copper silver selenide [*12040-91-4*],
      21:690
Copper telluride [*12181-15-6*], *23*:787, 801
Copper–tellurium alloys, *23*:802
Copper–tin alloys, *24*:117
   as shape-memory alloys, *21*:964
Copper-ware
   sulfamic acid cleaner for, *23*:129
Copper wires
   solders for, *22*:487
Copper–zinc alloys
   as shape-memory alloys, *21*:964
Coral reefs
   thorium dating of, *24*:70
Core competencies, *21*:270
Corfam, *24*:696
Cork binders
   sorbitol in, *23*:112

Corn
  centrifugal separation of, *21*:870
  extraction using SCFs, *23*:467
  starch from, *22*:699
  wet-milling, *22*:705
Corn cobs, *23*:96
Corning Code 7740
  attachment to vitreous silica, *21*:1041
Corning Code 7940, *21*:1034, 1049
  spectral normal emissivity, *21*:1057
Corn starch, *22*:705
  dextrose from, *23*:586
Corn sugar
  sorbitol from, *23*:105
Corn sweeteners
  syrup, *23*:582
Corn syrups [*8029-43-4*], *22*:699; *23*:39,
    582, 597
  use of sulfur dioxide in mfg, *23*:310
Coronite, *24*:404
Corresponding states theorem, *23*:1012
Corrosion
  analysis by Mössbauer spectroscopy,
    22:656
  of carbon steel in steam prdn, *22*:742
  in hydrothermal processing of wastes,
    22:760
  inhibition, *23*:212
  prevention for tanks, *23*:651
  prevention in acid manufacturing,
    23:389
  preventors in cooling towers, *22*:217
  protective coatings, *23*:1049
  rate in sulfamic acid, *23*:122
  rates for vitreous silica, *21*:1042
  resistance of steels, *22*:815
  of steam system components, *22*:727
  during sterilization, *22*:845
  sulfur, *23*:269
  in tanks and pressure vessels, *23*:637
  of tantalum, *23*:671
  by thiosulfates, *24*:56
Corrosion control
  silicates for, *22*:22
Corrosion inhibitors
  silicate–siliconate mixtures as, *22*:147
  sodium nitrite for, *22*:394
  strontium compounds as, *22*:953
  sulfonated matls for, *23*:162
  thioglycolic acid as, *24*:16
Corrosiveness
  of sulfuric acid solutions, *23*:393

Cor-Ten A, *22*:815
Corticosteroids, *22*:852
  as veterinary drugs, *24*:832
Corticosterone [*50-22-6*], *22*:860; *24*:832
Cortisol [*50-23-7*], *22*:860, 883; *24*:504, 832
Cortisone [*53-06-5*], *22*:860; *24*:832
  selenium in synthesis of, *21*:713
Corundum [*12252-63-0*], *21*:53
Corynebacterium sp.
  antimicrobial agent for, *24*:828
Cosmetics
  regulation of, *21*:177
  rheological measurements, *21*:427
  silica gels in, *21*:1023
  silk in, *22*:162
  soaps in, *22*:324
  sulfonic acid-derived dyes in, *23*:206
  sulfonic acids in, *23*:205
  talc application, *23*:613
  thioglycolic acid in, *24*:12
  triarylmethane dyes for, *24*:565
  use of hexitols in, *23*:107
Cottage cheese
  sorbates in, *22*:582
Cotton, *22*:593
  mechanical properties of, *22*:160
  sodium nitrate as fertilizer for, *22*:392
  sulfonic acid derived dye for, *23*:207
  sulfur dyes for, *23*:360
  in tire cord, *24*:162
  tire cords, *24*:162
  titanated inks for, *24*:329
Cotton–polyester blends, *23*:901
Cottonseed, *22*:591
  extraction using SCFs, *23*:467
Cottonseed oil
  specific gravity, *23*:625
Couette viscometers, *21*:394
Cough drops
  sugar alcohols in, *23*:108
4-Coumarate-CoA ligase, *23*:751
Coumarin
  additives for refractory coatings, *21*:110
Coupling agents, *22*:149
  sulfonated aromatics as, *23*:159
Coupling sugar, *23*:7
CP-88,818 [*99759-19-0*], *22*:907
CP-148,623 [*150332-35-7*], *22*:907
CPH-53-N
  fatty acid surfactant, *23*:518
CPSC. See *Consumer Product Safety
  Commission.*

C-reactive protein, 24:499
Cream
  centrifugal separation of, 21:859
  tea, 23:761
Creel calendering
  for tire use, 24:168
Creep
  in tire cord, 24:165
Creep experiments, 21:406
Creosote
  sampling standards for, 21:628
Crepes, 21:565
Cresol [1319-77-3]
  regulatory level, 21:160
m-Cresol [108-39-4]
  regulatory level, 21:160
  as solvent, 22:548
o-Cresol [95-48-7]
  regulatory level, 21:160
p-Cresol[106-44-5]
  regulatory level, 21:160
Cresylic acid [1319-77-3]
  sampling standards for, 21:628
  as solvent, 22:548
Cretinism
  role of the thyroid, 24:91
Crisping agents
  sulfonated aromatics as, 23:159
Cristobalite [14464-46-1], 21:981, 1006,
    1077
  in silica refractories, 21:83
Critical micelle concentration, 23:488
  in soap, 22:298
Critical pigment volume concentration
  of vinyl acetate paints, 24:973
Crookesite [12414-86-7], 23:952
Crospovidones, 24:1078
Cross equation, 21:351
Cross-linked starches, 22:714
Cross-linking
  of poly(vinyl acetals), 24:928
Cross-linking agents, 21:469
  in silicone sealant, 21:655
  silicon esters as, 22:80
Crotonaldehyde [4170-30-3]
  sorbic acid from, 22:575
Crown ethers
  alkali metals with, 22:331
Crucibles
  refractories for, 21:84
  refractory coatings for, 21:111

Crude oil, 23:718
  rheological measurements, 21:427
  supercritical fluid extraction, 23:465
Crude tall oil (CTO), 21:293; 23:616
Crumb rubber
  from scrap tires, 21:31
Crushing
  for size reduction, 22:283
Cryoelectronics
  silicon-based semiconductors, 21:744
Cryogenic process
  for grinding scrap tires, 21:34
Cryogenics
  solders for, 22:487
Cryptands
  alkali metals with, 22:331
Cryptococcosis, 24:829
Cryptococcus neoformans
  inhibited by sorbates, 22:580
Cryptococcus sp.
  inhibited by sorbates, 22:580
Cryptococcus terreus
  inhibited by sorbates, 22:579
Crystal growth experiment
  in reduced gravity, 22:621
Crystal hypo, 24:57
Crystalline fructose, 23:25
Crystalline silicon
  amorphous silicon contrasted, 21:757
Crystallization
  in dextrose, 23:591
  polymer by supercritical fluids, 23:460
  pure silicon, 21:1092
  role in separation process synthesis,
    21:959
  rubber, 21:570
Crystallization, fractional
  sodium chloride prdn, 22:365
Crystal violet [548-62-9], 24:552, 827
Crystex, 23:257
CSPE. See Chlorosulfonated polyethylene.
CST. See Catalyst Stabilization
    Technology.
CTGs. See Control Technique Guideline
    documents.
CTO. See Crude tall oil.
Cubes
  of sugar, 23:41
Cubic boron nitride
  as tool material, 24:444
Cumene [98-82-8]
  as HAP compound, 22:532

Cunninghamella echinulata
  inhibited by sorbates, *22*:579
Cup lump grades
  of rubber, *21*:571
Cupric chloride [*7447-39-4*], *24*:856
Cuprimyxin [*28069-65-0*], *24*:829
Cuprodescloizite [*12325-36-9*], *24*:783
Cuprous chloride [*7758-89-6*], *24*:853
Cuprous iodide
  in table salt, *22*:373
Cup viscometers, *21*:379
Curculin [*151404-13-6*], *23*:576
Curd
  from soap, *22*:297
Cure characteristics
  of rubber, *21*:571
Cured silicone LIM, *22*:113
Curie temperature
  for ferrite, *22*:793
Curing
  diorganotin catalysts for, *24*:146
Curing agents
  thioglycolic acids as, *24*:13
Current Good Manufacturing Practice
  (cGMP), *21*:172
Curvularia trifolii
  inhibited by sorbates, *22*:579
Cutlery
  steel for, *22*:822
Cutter
  insect repellent, *21*:242
Cutting
  for size reduction, *22*:283
Cutting mills, *22*:293
Cutting tools
  titanium compounds in, *24*:232
Cuvettes
  of vitreous silica, *21*:1065
CVM. See *Center for Veterinary Medicine.*
CWA. See *Clean Water Act.*
Cyanagel, *22*:456
Cyanazine [*21725-46-2*], *22*:422
Cyanide
  thiosulfate as antidote, *24*:55
Cyanide detoxification, *23*:312
Cyanides
  silver, *22*:169
  silver complexes of, *22*:185
Cyanidin [*528-58-5*], *23*:756
Cyanoethyltrimethoxysilane
  critical surface tension of, *22*:148

Cyanogen
  interference in atomic emission
    spectroscopy, *22*:647
Cyclamate [*100-88-9*]
  compared to sucrose, *23*:4
Cyclamic acid [*100-88-9*], *23*:566
Cyclic siloxanes, *22*:93
Cycloate
  thiols in, *24*:29
Cyclobuxine-D [*2241-90-9*], *22*:865
β-Cyclodextrin, *24*:500
Cyclohexane [*110-82-7*]
  dielectric constants of, *22*:759
  as solvent, *22*:536
trans-Cyclohexane-1,4-diisocyanate
    [*2556-36-7*], *24*:708
1,4-Cyclohexanedimethanol divinyl ether
    [*17351-75-6*], *24*:1066
Cyclohexanethiol [*1569-69-3*], *24*:21
Cyclohexanol [*108-93-0*]
  as solvent, *22*:536
Cyclohexanone [*108-94-1*]
  as solvent, *22*:546
1-(3-Cyclohexen-1-ylcarbonyl)hexahydro-
    1*H*-azepine [*52736-59-1*]
  as repellant, *21*:249
1-(3-Cyclohexen-1-ylcarbonyl)piperidine
    [*52736-58-0*]
  as repellant, *21*:249
Cyclohexylamine [*108-91-8*], *23*:566
  as solvent, *22*:540
  water additive of steam prdn, *22*:740
N-Cyclohexyl-2-benzothiazolesulfenamide
    (CBTS) [*95-33-0*], *21*:465
1-(Cyclohexylcarbonyl)-hexahydro-1*H*-
    azepine [*68571-09-5*]
  as repellant, *21*:249
Cyclohexylmethyldimethoxysilane
    [*17865-32-6*], *22*:147
N-(Cyclohexylthio)phthalimide [*17796-82-
    6*]
  as rubber chemical, *21*:473
Cyclone boilers
  refractories for, *21*:84
Cyclones, *21*:885
Cyclopentadienylthallium [*34822-90-7*],
    23:957; *24*:317
Cyclopentadienyltitanium compounds,
    24:312
Cyclophosphamide [*50-18-0*], *24*:836
Cyclosiloxanes, *22*:91
Cyclosporine [*59865-13-3*], *24*:837

Cyclotetrasiloxane D_1
  vibrational peaks of, 22:516
Cyclotrisiloxane D_2
  vibrational peaks of, 22:516
Cylinderite [59858-98-9], 24:105
Cylindrite [12294-05-0], 24:123
p-Cymene [25155-15-1], 23:834
Cymet, 23:246
Cysteine [52-90-4], 24:20
  in silk, 22:156
Cystic fibrosis, 24:506
Cystine [923-32-0], 24:2
  in soybeans and other oilseeds, 22:596
Cytosine arabinoside hydrochloride
  [69-74-9], 24:836
Czochralski technique, 23:677, 783

D

2,4-D [94-75-7], 22:422
Dabco, 24:699
Dacthal [1861-32-1], 22:424; 24:510
Daidzein [486-66-8]
  in soybeans, 22:598
Dairy equipment
  sulfamic acid cleaner for, 23:129
Dairy products
  dextrose for, 23:593
  high fructose syrup in, 23:597
  sorbates in, 22:582
Dairy wheys
  use of ultrafiltration, 24:621
Dammar gum [9000-16-2], 21:297
Dammar resins [9000-16-2], 21:297
Damping
  of alloys, 21:964
D'ansite, 22:404
DAPEX process
  for uranium recovery, 24:650
Darapskite, 22:386
Darcy's law, 24:606
Darrieus machine
  for wind energy technology, 22:467
Databases, 23:456
  for infrared spectroscopy, 22:640
  use in technical service, 23:779
Dates, 23:96
  trace analysis of, 24:515
Dating
  thorium isotopes for, 24:70
Davanite, 24:264

Davidite [12173-20-5], 24:783
  magnetic intensity, 21:876
Daxad
  naphthalene sulfonate surfactant,
    23:497
DBU, 24:699
D&C Blue No. 9, 23:549
DCFET. See Doped channel field-effect
  transistor.
D&C Green No. 5, 23:550
DDBSA
  alkylbenzensulfonate surfactant, 23:496
DDT [50-29-3], 22:421
DEAE-cellulose–titanium dioxide–
    polystyrene
  support for enzymes, 23:595
Debaryomyces membranaefaciens
  inhibited by sorbates, 22:580
Debaryomyces spp.
  inhibited by sorbates, 22:580
Deborah number, 21:369
Decaffeination
  of tea, 23:762
Decamethylcyclopentasiloxane [541-02-6],
    22:103
Decamethyltetrasiloxane [141-62-8],
    22:103
1-Decanethiol [143-10-2], 24:21
Decarburization, 24:424
DECHEMA, 23:456
Dechlorination
  water treatment for steam prdn, 22:745
Decontamination, 22:848
n-Decyl hydrogen sulfate [142-98-3],
    23:410
Deer repellents, 21:257
DEET, 21:236
DEF
  thiols in, 24:30
Defect states
  in amorphous silicon, 21:750
Deflocculants
  silicates as, 22:24
Defoamers
  precipitated silica as, 21:1025
  silicas in, 21:1002
Degrees of freedom, 23:1024
Dehydroabietic acid [1740-19-8], 21:294
7-Dehydrocholesterol [434-16-2], 22:856
Dehydroepiandrosterone (DHEA)
  [53-43-0], 22:859
Dehydroepiandrosterone acetate [853-23-
  6], 22:875, 880

(±)-9,11-Dehydroesterone methyl ether
[*1670-49-1*], *22*:889
Dehydrogenation
of alcohols, *21*:343
Dehydronerolidol [*2387-68-0*], *23*:870
5-Dehydroshikimate reductase, *23*:752
Dehyquart
surfactant, *23*:529
Deicer
sodium chloride as, *22*:373
Deinking
of paper, *21*:14
Deinking surfactants
in paper recycling, *21*:16
Deklene, *23*:543
Delayed coking, *23*:738
Delayed cure
for textile finishing, *23*:900
Delivery receipt, *24*:532
Delphinidin [*528-53-0*], *23*:756
Deltamethrin [*52918-63-5*]
as bird repellent, *21*:256
Demeton
thiols in, *24*:29
Dendrite growth kinetics, *22*:625
Densification
role in sol–gel technology, *22*:74, 518
Density gradient
separations, *21*:853
Dental implants
tantalum for, *23*:671
Dental materials
shape-memory alloys for, *21*:973
silver–mercury amalgams, *22*:177
from sol–gel technology, *22*:526
tin alloys in, *24*:121
Dentifrices
polyols in, *23*:106
saccharin in, *23*:566
silica gels in, *21*:1023
Deodorant sticks, *22*:324
7-Deoxycholic acid [*83-44-3*], *22*:856
Deoxynivalenol, *22*:599
Deoxyribonucleic acid
trace analysis, *24*:516
Department of Transportation, *21*:165
Dephosphorization
role in steelmaking, *22*:781
Depitching, *23*:620
Depolymerized scrap rubber, *21*:31
Depression
stimulants for, *22*:938

Derivatization
use of silylating agents in, *22*:143
Dermalon, *23*:543
Dermatitis
selenium cmpds for, *21*:713
Desalination
by reverse osmosis, *21*:324
steam heat for, *22*:758
Descaling agent
sodium hydride as, *22*:345
Descloizite [*19004-61-6*], *24*:783
Desiccating agent, *24*:258
Desipramine [*50-47-5*], *22*:942
Desliming, *21*:906
Desmodur L, *24*:716
Desugarization
of molasses, *23*:58
Desulfurization
flue gas, *23*:447
role in steelmaking, *22*:781
sodium hydrosulfide from, *22*:412
Detectivity
in optical spectroscopy, *22*:631
Detectors
for optical spectroscopy, *22*:635
rubidium compounds as, *21*:598
Detergency
fatty acid esters of hexitols, *23*:108
Detergent builders
sucrose carboxylates as, *23*:73
Detergents, *23*:478
linear alkylbenzenesulfonic acids in,
23:205
precipitated silica in, *21*:1025
sampling standards for, *21*:628
silicates in, *22*:21
use of sodium sulfate for, *22*:410
use of sulfur for, *23*:251
Detersive systems, *23*:477
Detinning, *24*:111
Detonimide, *24*:834
DETU. See *N,N-Diethylthiourea*.
Deuterium [*16873-17-9*]
as component of steam, *22*:721
lamp for optical spectroscopy, *22*:635,
646
Deuterium selenide [*13536-95-3*], *21*:697
Device physics, *21*:732
Dewar flasks
chemical silvering of, *23*:1074
Dewatering, *21*:906
centrifuges for, *21*:847

Dexamethasone [50-02-2], 22:905; 24:832
Dexon, 23:542
Dexon II, 23:543
Dexon "S", 23:543
Dextrans, 23:8
    in beet sugar, 23:55
Dextrose [50-99-7], 23:582
    in corn syrup, 23:598
Dextrose hydrate [16824-90-1], 23:583
D-gun, 21:90
DHEA. See Dehydroepiandrosterone.
DHT. See Dihydrotestosterone.
Diacetato 2-methoxy-2-phenylethane
    tallium(III) [37011-27-1], 23:959
Diacetone alcohol [123-42-2]
    as solvent, 22:546
Diacetoxyscirpenol, 22:599
Diafiltration, 24:603, 616
Dial, 22:323
Dialkyl sulfates, 23:417
Diammonium imidodisulfonate [13597-
    84-1], 23:124
Diamond
    for optical spectroscopy, 22:635
    thin films of, 23:1066
    as tool material, 24:436
1,4:3,6-Dianhydro-D-glucitol, 23:100
1,3:2,5-Dianhydroxylitol, 23:100
1,1-Di-p-anisylethane [10543-21-2], 24:856
Diapirs, 22:355
Diatomaceous earth [7631-86-9], 21:992,
    1028; 23:105; 24:862
    as catalyst support, 23:389
6,6'-Diazido-6,6'-dideoxysucrose, 23:71
Diazinon [333-41-5]
    as insect repellent, 21:253
20,25-Diazocholesterol dihydrochloride
    [1249-84-9]
    as bird repellent, 21:256
Dibasic ester
    as solvent, 22:544
Dibenzyltin dichloride [3002-01-5], 24:138
Diborane
    dopant for pure silicon, 21:1095
1,2-Dibromo-3-chloropropane [76-12-8],
    22:423
1,2-Dibromoethane [106-93-4]
    as solvent, 22:538
cis-2,5-Dibromosulfolane [30186-52-8],
    23:135
trans-2,5-Dibromosulfolane [30186-54-0],
    23:135

2,5-Dibromothiophene [3141-27-3], 24:43
Di-t-butoxydiacetoxysilane [13170-23-5],
    22:72
Dibutylamine [111-92-2]
    as solvent, 22:540
Di-n-butyl methylene bisthioglycolate
    [14338-82-0], 24:3, 7
Di-t-butyl peroxide [110-05-4]
    silicone peroxide cure, 22:107
2,5-Di(t-butylperoxy)2,5-dimethylhexane
    [78-63-7]
    silicone peroxide cure, 22:108
Dibutyl phthalate [84-74-2]
    as mosquito repellent, 21:244
Di(n-butyl) sulfate [625-22-9], 23:410
Di(n-butyl) sulfite [626-85-7], 23:410
Di-t-butylsulfoxide [2211-92-9], 23:221
N,N-Di-n-butylthiourea [109-46-6], 21:467
Dibutyltin bis(isooctyl thioglycolate)
    [25168-24-5], 24:13
Dibutyltin diacetate [1067-33-0], 24:145
Dibutyltin dilaurate [77-58-7], 24:146
Dibutylxanthogen [105-77-1], 21:468
Dibutylxanthogen polysulfide (XPS),
    21:468
Dicamba [1918-00-9], 22:422
Dichloroacetic acid [79-43-6], 24:4
3,4-Dichloroaniline [95-76-1], 22:427
Dichlorobenzene
    rejection by reverse osmosis membrane,
    21:319
o-Dichlorobenzene [95-50-1]
    as solvent, 22:538
1,4-Dichlorobenzene [106-46-7]
    regulatory level, 21:160
Dichloroethane [1300-21-6], 24:852
1,1-Dichloroethane [75-34-3], 24:855
    as HAP compound, 22:532
1,2-Dichloroethane [107-06-2], 24:852
    as HAP compound, 22:532
    regulatory level, 21:160
    as solvent, 22:538
cis-1,2-Dichloroethylene [156-59-2], 24:865
trans-1,2-Dichloroethylene [156-60-5],
    24:865
1,1-Dichloroethylene [75-35-4], 24:865,
    882
    regulatory level, 21:160
β,β-Dichloroethyl phenyl sulfone
    [3123-10-2], 24:856
Di-2-chloroethyl sulfate [5411-48-3],
    23:410

Dichlorofluanid [1085-98-9], 23:275
1,1-Dichloro-3-methylbutane [625-66-1], 24:856
Dichloromethyl sulfate [73455-05-7], 23:410
2,4-Dichlorophenol [120-83-2], 22:427
2,4-Dichlorophenoxy [94-75-7]
  regulatory level, 21:160
(2,4-Dichlorophenoxy)phenol, 24:510
Dichlorophenylthallium(III) [19628-33-2], 23:958
2,5-Dichlorophenylthioglycolic acid [6274-27-7], 24:16
2,3-Dichloro-1-propanol [616-23-9], 24:856
3,3-Dichloro-1-propanol [83682-72-8], 24:856
Dichloropropene, 22:422
Di-3-chloropropyl sulfite [83929-99-1], 23:410
3,6-Dichlorosalicylic acid [3401-80-7], 22:427
Dichlorosilane [4109-96-0], 21:1087; 22:36
N,N-Dichlorosulfamic acid [17085-87-9], 23:123
3,4-Dichlorosulfolane [3001-57-8], 23:135
Dichlorotitanium diacetate [4644-35-3], 24:297
N,N'-Dicinnamylidene-1,6-hexane diamine [140-73-8]
  cross-linking agent, 21:470
Dicumyl peroxide [80-43-3]
  cross-linking agent, 21:470
  rubber curing, 21:573
  silicone peroxide cure, 22:108
N-Dicyclohexyl-2-benzothiazole-sulfenamide [4979-32-2], 21:465
Dicyclohexylmethane diisocyanate [5124-30-1], 24:708
Dicyclopentyl dimethoxysilane [126990-35-0], 22:147
Di(n-decyl) sulfate [66186-16-1], 23:410
Di(3,4-dichlorobenzoyl) peroxide [133-14-2]
  silicone peroxide cure, 22:108
Dieldrin [60-57-1], 22:421
Dielectric insulators, 21:808
Dielectrics
  from silicon ester chemical vapor deposition, 22:80
Diels-Alder addition
  of sulfur compounds, 23:272
Diels hydrocarbon [549-88-2], 22:852

Diesel fuel
  removal from soil using SCF, 23:466
Dietary supplement
  regulatory agencies, 21:176
Diethanolamine [111-42-2]
  for hydrogen sulfide removal, 23:437
  as solvent, 22:540
  water additive of steam prdn, 22:740
Diethylamine [109-89-7]
  as solvent, 22:540
m-Diethylbenzene [141-93-5], 22:990
p-Diethylbenzene [105-05-5], 22:990
Diethylbenzenes, 22:960
Diethyl(n-butyl) sulfate [5867-95-8], 23:410
Diethylcarbamazine [98-89-1], 24:831
Diethyldithiocarbamate
  determination of silver, 22:187
Diethylene glycol [111-46-6]
  as solvent, 22:546
Diethylene glycol monobutyl ether [112-34-5]
  as solvent, 22:544
Diethylene glycol monobutyl ether acetate [124-17-4]
  as solvent, 22:544
Diethylene glycol monoethyl ether [111-90-0]
  as solvent, 22:544
Diethylene glycol monoethyl ether acetate [112-15-2]
  as solvent, 22:542
Diethylene glycol monomethyl ether [111-77-3]
  as solvent, 22:544
Diethylene glycol monomethyl ether acetate [629-38-9]
  as solvent, 22:542
Diethylene glycol monostearate [106-11-6]
  as release agent, 21:210
Diethylenepentaaminetetraacetic acid (DPTA)
  in pulping, 21:14
Diethylenetriamine [111-40-0]
  as solvent, 22:540
Diethyl ether [60-29-7], 24:853
N,N-Diethyl-3-methylbenzamide, 21:236
Diethylsilane [542-91-6], 22:38
Diethylstilbestrol [56-53-1], 24:833
Diethyl succinate [123-25-1], 22:1076
Diethyl sulfate [64-67-5], 23:410

Diethyl sulfite [623-81-4], 23:410
N,N-Diethylthiourea (DETU) [105-55-5], 21:467
N,N-Diethyl-m-toluamide [134-62-3], 21:236
2,2-Difluoro-4-chloro-1,3-dioxolane [162970-83-4], 24:856
1,1-Difluoroethane [75-37-6], 24:855
Difolatan, 23:291
Difunctional initiators
    for styrene polymerization, 22:1039
Digitoxigenin [143-62-4], 22:866
Diglycolamine [929-06-6]
    for hydrogen sulfide removal, 23:437
Digoxigenin, 24:503
Digoxin, 24:503
Dihydroflavonol 4-reductase, 23:751
Dihydrolinalool [18479-49-7], 23:856
Dihydroterpineol [498-81-7], 23:855
Dihydroterpinyl acetate [80-25-1], 23:855
Dihydrotestosterone (DHT) [521-18-6], 22:859
4,4'-Dihydroxydiphenyl sulfone [80-09-1], 23:141
1,25-Dihydroxyvitamin D₃ [32511-63-0], 22:856
Diiodomethyl p-tolylsulfone [20018-09-1], 23:141
Diisobutyl carbinol [108-82-7]
    as solvent, 22:536
Diisobutyl ketone [106-83-8]
    as solvent, 22:546
Diisocyanate
    polyurethanes, 24:704
Diisocyanate prepolymer
    urethane sealants, 21:658
1,6-Diisocyanato-2,2,4,4-tetramethyl-hexane [83748-30-5], 24:707
1,6-Diisocyanato-2,4,4-trimethylhexane [15646-96-5], 24:708
Diisopropanolamine [110-97-4], 23:440
    as solvent, 22:540
Diisopropylamine [108-18-9]
    as solvent, 22:540
Diisopropyl succinate [924-88-9], 22:1076
Diisopropyl sulfate [2973-10-6], 23:410
Diisopropyl sulfite [4773-13-1], 23:410
Dilauryl thiodipropionate [123-28-4], 24:7
Dimensional stability
    of tire cord, 24:165
p-Dimethylamino-benzlidenerhodanine
    silver detection, 22:187

Dimethylaminoethylmethacrylate
    VP copolymerization, 24:1090
11η-(4-Dimethylaminophenyl)-17α-hydroxy-17η-(3-hydroxypropyl)-13α-methyl-4,9-gonadien-3-one, 22:903
Dimethylammonium hydrogen
    isophthalate, 21:468
3,5-Dimethylbenzenesulfonic acid [18023-22-8], 23:203
2,6-Dimethyl crystal violet [117071-61-1], 24:554
Dimethylcyclohexylamine [98-96-2], 24:699
2,5-Dimethyl-2,5-di(benzoylperoxy) hexane [2618-77-1]
    cross-linking agent, 21:470
2,5-Dimethyl-2,5-di(t-butylperoxy) hexane [78-63-7]
    cross-linking agent, 21:470
2,5-Dimethyl-2,5-di(t-butylperoxy) hexyne [1068-27-5]
    cross-linking agent, 21:470
Dimethyldichlorosilane [75-78-5], 22:83
    silylating agent, 22:144
Dimethylethanolamine [108-01-0]
    as solvent, 22:540
N,N-Dimethylformamide [68-12-2], 24:858
    as HAP compound, 22:532
    as solvent, 22:548
Dimethyloctadecyl-3-trimethoxy-silylpropylammonium chloride [27668-52-6], 22:150
    for liquid crystals, 22:149
1,3-Dimethylolethyleneurea
    in textile finishing, 23:900
N,N-Dimethyl-p-phenylenediamine
    for hydrogen sulfide detection, 23:282
Dimethyl phthalate [131-11-3]
    as mosquito repellent, 21:241
Dimethyl pyrosulfate [10506-59-9], 23:423
Dimethyl silicone [63148-62-9], 22:110
Dimethyl succinate [106-65-0], 22:1076
Dimethyl sulfate [77-78-1], 23:410
Dimethyl sulfite [616-42-2], 23:410
Dimethyl sulfone [67-71-0], 23:141
Dimethyl sulfoxide [67-68-5], 23:217; 24:854
Dimethylsulfoxonium methylide [5367-24-8], 23:223
Dimethyltelluride [593-80-6], 23:800

Dimethyltellurium dichloride [24383-90-3], 23:800
Dimethyl telluroketone [83270-40-0], 23:800
Dimethyl tellurone [83270-39-7], 23:800
Dimethyl tetrachloroterephthalate [1861-32-1], 22:424
Dimethyl thiodipropionate [4131-74-2], 24:7
Dimsyl ion, 23:222
Dingot metal, 24:655
4,4'-Dinitro-2,2'-dinitrostilbenedisulfonic acid [128-42-7]
    stilbene dye precursor, 22:922
Dinitrophenol
    rejection by reverse osmosis membrane, 21:319
Dinitroso-pentamethylenetetramine [101-25-7]
    as blowing agent, 21:479
2,4-Dinitrotoluene [121-14-2]
    regulatory level, 21:160
Dinonylnaphthalene sulfonic acid, 23:206
Dinoseb [88-85-7], 22:425
Dioctyltin bis(isooctylmercaptoacetate), 24:145
Dioctyltin maleate [16091-18-2], 24:145
Diodes, 21:732
    pure silicon for, 21:1102
Dioptase [15606-25-4], 22:3
Diosgenin [512-04-9], 22:853, 862
Dioxane
    formation during sulfation, 23:171
1,4-Dioxane [123-91-1]
    as HAP compound, 22:532
    as solvent, 22:548
2,5-p-Dioxanone, 23:548
Dioxin, 24:491
Dipentamethylenethiuram disulfide [94-37-1], 21:467
Dipentamethylenethiuram hexasulfide [971-15-3], 21:467
Dipentamethylenethiuram monosulfide [725-32-6], 21:467
Dipentene
    sampling standards for, 21:628
O-Diphenol, 23:751
Diphenyl ditelluride [32294-60-3], 23:800
1,1-Diphenylethane [612-00-0], 24:856
Diphenylguanidine [102-06-7], 21:467
P,P'-Diphenylmethylenediphosphinic acid, 24:294

Diphenylnaphthylmethanes
    as dyes, 24:551
Diphenyl oxide [101-84-8]
    as solvent, 22:548
Diphenyl sulfate [4074-56-0], 23:410
Diphenyl sulfite [4773-12-0], 23:410
Diphenyl sulfoxide [945-57-7], 23:225
Diphenyl telluride [1202-36-4], 23:800
Diphenyl telluroxide [51786-98-2], 23:800
N,N'-Diphenylthiourea (DPTU) [102-08-9], 21:467
Dipole moment
    role in optical spectroscopy, 22:648
Dipped goods, 21:584
DIPPERS, 23:456
Dipropetryn
    thiols in, 24:29
N,N-Dipropylamino-3-cyclohexene-carboxamide [68571-08-4], 21:246
N,N-Dipropylcyclohexanecarboxamide [67013-94-9]
    as repellant, 21:249
Dipropylene glycol [110-98-5]
    as solvent, 22:546
Dipropylene glycol monomethyl ether [34590-94-8]
    as solvent, 22:544
Dipropylene glycol monomethyl ether acetate [88917-22-0]
    as solvent, 22:544
Di(n-propyl) sulfate [598-05-0], 23:410
Di(n-propyl) sulfite [623-98-3], 23:410
Dipyridyl
    silver complexes of, 22:186
Dipyrone [5907-38-0], 24:832
Direct Brown 29, 22:923
Direct Orange 15 [1325-35-5], 22:926
Direct Orange 28, 22:923
Direct Orange 34 [32651-66-4], 22:923
Direct process
    for silicones, 22:84
Direct Yellow 4 [3051-11-4], 22:924, 925
Direct Yellow 6, 22:926
Direct Yellow 11 [1325-37-7], 22:922, 926
Direct Yellow 12 [2870-32-8], 22:926
Direct Yellow 19, 22:923
Direct Yellow 106 [12222-60-5], 22:924
Dirhenium decacarbonyl [14285-68-8], 21:343
Disc Tube, 21:306
1,4-Diselenane [1538-41-6], 21:703
Diselenium dichloride [10025-68-0, 21317-32-8], 21:687

Dishwashers
  sulfamic acid cleaner for, 23:129
Disilane [1590-87-0], 22:38; 24:854
Disilanylphosphine, 22:43
Disilver selenide [1302-09-6], 21:690
Disilylphosphine [14616-42-3], 22:43
Disinfectants, 22:847
  silver cmpds as, 22:167, 192
Disk memories
  refractory coatings for, 21:113
Disodium acetylide, 22:332
Disodium diselenide [39775-49-0], 21:687
Disodium selenide [1313-85-5], 21:687
Dispersants
  lignosulfates as, 23:205
  for pesticides, 23:169
  for tall oil recovery, 23:619
  thioglycolic acid as, 24:13
Dispersions
  organic matls in carbon dioxide, 23:462
  in paper recycling, 21:17
Display cases
  refrigeration in, 21:128
Distearyl thiodipropionate [693-36-7], 24:7
Distillation
  combined with reverse osmosis, 21:325
  role in separation process synthesis, 21:925
  simultaneous heat and mass transfer, 22:198
  steam heat for, 22:758
  of tall oil, 23:619
Distillation region diagrams
  role in separation process synthesis, 21:929
Distilled tall oil (DTO), 23:620
Distillers' grains
  use of sorbates, 22:584
Distress calls
  as bird repellents, 21:254
Distressing stains, 22:697
2,3-Disulfosuccinic acid [54060-35-4], 22:1080
Disulfoton
  thiols in, 24:29
Disulfuric acid, 23:373
Diterpene, 23:833
Di(n-tetradecyl) sulfate [66186-19-4], 23:410
Dithiobisbenzanilide [135-57-9]
  in rubber processing, 21:478
Dithiocarbamates
  vulcanization accelerator, 21:462

Dithiodiglycolic acid [505-73-7], 24:2, 7
4,4′-Dithiodimorpholine [103-34-4]
  cross-linking agent, 21:470
Dithiodipropionic acid [1119-62-6], 24:7, 16
1,4-Dithioglycolide [4835-42-6], 24:2
Dithionite, 23:317
Dithionous acid [15959-26-9], 23:312
Dithiophosphates
  vulcanization accelerator, 21:463
Dithiosalicylic acid [527-89-9], 21:623
1,1-Ditolylethane [29036-13-3], 24:856
Di-ortho-tolylguanidine (DOTG) [97-39-2], 21:467
6,6′-Di-O-tosylsucrose, 23:69
Di(trifluoroacetato)phenyltallium(III) [23586-54-1], 23:958
Di(triphenylsilane) [1450-23-3], 22:55
6,6′-Di-O-tritylsucrose [35674-15-8], 23:64
Dittus-Boelter equation, 22:199
Ditungsten boride [12007-09-9], 24:597
Ditungsten carbide [12070-13-2], 24:597
Ditungsten nitride [12033-72-6], 24:597
Ditungsten pentaboride [12007-98-6], 24:597
Ditungsten trisilicide [12138-30-6], 24:598
Divergan, 24:1094
Divinylbenzene [1321-74-0], 22:988
Divinylbenzene copolymers, 22:1048
Divinylcopper lithium [22903-99-7], 24:854
Divinylsilane [18142-56-8], 22:42
Divinyl sulfide [627-51-0], 24:854
DLVO theory
  of colloid stability, 21:672
Docosylamine
  thin films of, 23:1086
Dodecamethylcyclohexasiloxane [540-97-6], 22:103
Dodecamethylpentasiloxane [141-63-9], 22:103
tert-Dodecanethiol [25103-58-6], 24:21
1-Dodecanethiol [112-55-0], 24:21
Dodecanoic acid
  as rubber chemical, 21:473
Dodecylbenzenesulfonic acid [27176-87-0], 23:198, 203
  siloxane catalyst, 22:93
n-Dodecyl hydrogen sulfate [151-41-7], 23:410
Dodecylthioacetic acid [13753-71-4], 24:2, 16

Dodecylthiopropionic acid [1462-52-8], 24:16
Doebner's violet [3442-83-9], 24:553
Dog repellents, 21:259
Dolomite [17069-72-6], 21:55; 22:767
  refractories from, 21:50
Donan dialysis
  sulfonic acid derivatives for, 23:169
Donnan exclusion
  reverse osmosis models, 21:308
Doors
  PVC use, 24:1040
Dopamine [51-61-6], 22:939
  trace analysis of, 24:507
Dopants, 21:803
  into thin films, 23:1066
Dopant sources
  in MOCVD, 21:774
Doped channel field-effect transistor
    (DCFET), 21:780
Doped semiconductors
  grown in microgravity, 22:623
Doppler broadening
  role in optical spectroscopy, 22:632
Doppler shifts
  of scattered light, 22:648
Dorlastan, 24:718
Dorzolamide [120279-96-1], 24:48
Dosimeters
  for sterilization, 22:839
DOTG. See Di-ortho-tolylguanidine.
DOTG salt [16971-82-7], 21:467
Double polarization
  in sugar analysis, 23:14
Dove, 22:323
Dowex, 24:498
Dowfax 2A1, 23:499
Dowfax 3B2, 23:499
Downs cell, 22:336
Doxapram [309-29-5], 22:935
Doxepin [1668-19-5], 22:942
Doxorubicin [23214-92-8], 24:836
DPTA. See Diethylenepenta-
    aminetetraacetic acid.
DPTU. See N,N'-Diphenylthiourea.
Draeger tubes, 23:282
Draglines, 23:732
Drains
  refractories for, 21:84
Draka, 24:710
DRAM cells, 21:741
Drilling muds
  lignosulfonates for, 23:169

Droloxifene [82413-20-5], 22:904
Droperidol [548-73-2], 24:835
Droplet sizes
  in sprays, 22:678
Droxifilcon-A, 24:1079
Drugs. See Pharmaceuticals.
Drum agglomerators, 22:230
Dry-cleaning
  role of titanates, 24:324
Dry-cleaning solvent, 22:569
Dry etching
  compound semiconductor processing,
    21:800
Drying
  of soap, 22:314
  of tea, 23:754
  use of SCFs, 23:465
Drying methods, 22:229
Drying oils
  Tetralinoleic alcohol titanate, 24:329
Dry rubber, 21:563
DTO. See Distilled tall oil.
Ductile iron, 23:392
Dulcin [150-69-6]
  potential sweetener, 23:573
Dulcitol [608-66-2], 23:95
Du Nouy tensiometer, 23:491
Duponol C
  surfactant, 23:501
Dupré equation, 23:484
Durable press garments, 23:893
Duranon Tick Repellent, 21:250
Durimet 20, 23:307, 393
Duriron, 23:394
Dutch twill
  centrifuge filters, 21:867
Dyeing
  of textiles, 23:888
Dyes
  in soap, 22:319
  sodium nitrite for, 22:394
  sulfamic acid in mfg, 23:129
  sulfonation of, 23:158
  sulfonic acid-derived, 23:206
  use of selenium, 21:715
  in wood stains, 22:692
Dykolite Brilliant Orange 3G
  sulfur dye, 23:343
Dynamic Flex Strip Adhesion
  adhesion in tire cord, 24:181
Dynamic light scattering
  for particle size measurement, 22:270

Dynamic viscometer, *21*:392
Dynamites
  sodium nitrate in, *22*:392
Dynasil, *21*:1049, 1058
Dynasil 1000, *21*:1034
DynaWave, *23*:391
Dynels
  membrane for ultrafiltration, *24*:604
Dyphylline [*479-18-5*], *22*:936
Dzhezkazganite, *21*:336

E

Earl Grey tea, *23*:759
East India resins [*9000-16-2*], *21*:297
EBEX, *22*:967
EBMax
  alkylation technology, *22*:966
Ebonex, *24*:234
Ebselen, *21*:713
EC 1.10.3.2, *23*:751
EC 1.11.1.7, *23*:752
EC 1.12.12.11, *23*:751
EC 1.14.11.9, *23*: 751
EC 1.14.18.1, *23*:751
EC 6.2.112, *23*:751
Ecdysone [*3604-87-3*], *22*:868
Ecdysteroids, *22*:852
E. coli. See *Escherichia coli.*
Econo Foam, *24*:710
ECPA. See *Electric Consumers Protection
  Act.*
ECR. See *Electron cyclotron resonance
  etching.*
Eddie Bauer Insect Formula, *21*:243
Edeleanu process, *23*:311
Edetic acid [*60-00-4*], *23*:570
Edman degradations, *22*:151
EDTA. See *Ethylenediaminetetraacetic
  acid.*
Effluents
  ftir analysis, *22*:642
Efflux viscometers, *21*:379
Eggs
  extraction of lipids, *23*:467
Eicosadienoic acid [*25448-01-5*], *23*:617
Eicosamethylnonasiloxane [*2652-13-3*],
  22:103
Einstein equation, *21*:362
EIS. See *Environmental impact statement.*
Elastomers, *21*:482
  for centrifuges, *21*:847

diorganotin catalysts for, *24*:146
polysulfide, *22*:417
precipitated silica in, *21*:1025
Electrical conductors
  thin films for, *23*:1071
Electrical connectors
  shape-memory alloy as, *21*:969
  use of tellurium alloys, *23*:802
Electrical contacts
  brazing filler metals for, *22*:490
Electrical generation
  scrap tire fuel, *21*:25
Electrical materials
  silica for, *21*:999
Rlectric Consumers Protection Act
  (ECPA), *21*:181
Electric furnace processes
  for steel, *22*:768
Electric insulation
  PVF resins, *24*:938
Electricity
  from steam, *22*:755
Electric power
  regulatory agencies, *21*:179
Electric vehicles
  sodium batteries for, *22*:340
Electroceramics
  titanium dioxide for, *24*:235
Electrochemical process
  for prdn of sodium dithionite, *23*:317
Electrochromatography
  for trace and residue analysis, *24*:507
Electrodeposition
  reduced gravity experiments, *22*:625
  of refractory coatings, *21*:91
  of tellurium, *23*:804
Electrodes
  silver–silver salt, *22*:190
  silylating agent on, *22*:149
  for steelmaking, *22*:771
Electrodialysis
  for sodium chloride prdn, *22*:366
Electroforming
  nickel sulfamate for, *23*:130
  of silver, *22*:169
Electroless plating, *23*:1072
  of refractory coatings, *21*:92
  of silver, *22*:176
  use of selenium, *21*:715
Electrolysis cell
  thin films from, *23*:1068
Electromagnetic interference, *24*:327

Electromagnetic spectrum, 22:629
Electron accelerators
  for sterilization, 22:845
Electron capture
  for trace and residue analysis, 24:500
Electron cyclotron resonance (ECR)
  etching, 21:800
Electronic devices
  compound semiconductors, 21:776
Electronic mail
  use in technical service, 23:780
Electronic materials
  sulfolane as solvent, 23:140
  tantalum as, 23:670
  thorium as, 24:71
  ultrapure water for, 21:326
Electron tubes
  rhenium alloys for cathodes, 21:340
Electrophoresis
  for refractory coatings, 21:101
  soybean proteins, 22:595
Electrophotography
  amorphous semiconductors in, 21:750
Electroplating
  brass, 23:208
  nickel sulfamate for, 23:130
  of rhenium, 21:341
  of silver, 22:169, 176
  silver cyanide for, 22:191
  tellurium as additive for, 23:804
  thin films for, 23:1068
  wastewater treatment for, 21:320
Electrorheological fluids, 21:365
Electroslag
  remelting of, 22:772
Electrostatic atomizer, 22:672
Electrotinning, 24:125
Electroultrafiltration, 24:614
Elisagram, 24:500
Elisa kit, 24:500
Elsorb process, 23:449
Emcol
  surfactant, 23:531
Emerest
  glycerol ester surfactant, 23:512
  polyoxyethylene surfactant, 23:514
Emerest 2350
  fatty acid surfactant, 23:518
Emerest 2355
  fatty acid surfactant, 23:518
Emerest 2381
  fatty acid surfactant, 23:518

Emergency call boxes
  solar energy for, 22:475
Emid
  diethanolamine surfactant, 23:520
Emid 6500
  surfactant, 23:520
Emission limits, 21:158
Emissivities
  role in temperature measurement,
    23:827
Emkal
  alkylbenzensulfonate surfactant, 23:496
Emmerich method
  reducing sugar test, 23:16
Emolliency, 23:480
Emsorb
  sorbitan surfactant, 23:516
Emsorb 2500
  surfactant, 23:515
Emsorb 2502
  surfactant, 23:515
Emsorb 2503
  surfactant, 23:515
Emsorb 2505
  surfactant, 23:515
Emsorb 2510
  surfactant, 23:515
Emsorb 2515
  surfactant, 23:515
Emulsifiers
  sorbitan fatty esters as, 23:110
  sulfonated matls for, 23:162
Emulsion polymerization
  of styrene–butadiene, 22:999
Emulsion steam drive, 23:729
Emulsion styrene–butadiene rubber
    (ESBR), 22:995
Enamels, 24:127
  as refractory coatings, 21:102
  sodium nitrate for prdn of, 22:393
Enamels, vitreous
  titanium dioxide in, 24:238
S-(−)-enantiomer [4276-74-8], 23:577
Enantiomers
  as sweeteners, 23:558
Encephalitis, 21:238
Endomycopsis ohmeri
  inhibited by sorbates, 22:580
Endosulfan [115-29-7]
  from thionyl chloride, 23:295
Endrin [72-20-8]
  regulatory level, 21:160

Energy conservation
  in thermodynamics, 23:985
Energy-dispersive x-ray fluorescence,
  22:655
Energy Policy Act (EPACT), 21:181
Enhanced oil recovery
  silicates in, 22:23
  sulfonated matls for, 23:162
  titanates for, 24:332
  use of SCFs, 23:465
Enhanced petroleum recovery
  sulfolane as cosurfactant, 23:140
Enhancement factors
  for supercritical systems, 23:463
Enprofylline [41078-02-8], 22:936
Enterobacter aerogenes
  inhibited by sorbates, 22:580
Entropy, 23:986
Entropy increase, 23:985
Environmental considerations
  mining processes, 23:740
Environmental degradation
  of polystyrene, 22:1033
Environmental impact statement (EIS),
  21:165
Environmental Planning and Community
  Right-to-Know Act (EPCRA), 21:165
Environmental protection, 21:150
Environmental Protection Agency, 21:165
Enviroseal, 22:79
Enzymatic methods
  in sugar analysis, 23:16
Enzyme-linked immunosorbent assay,
  24:514
Enzymes
  dextrose in prdn of, 23:592
  effect of SCFs as solvents, 23:469
  immobilization using silylating agents,
    22:151
  in soybeans and other oilseeds, 22:596
  use of ultrafiltration, 24:622
Enzyme technology
  applied to starch, 22:699
Enzyme treatment
  in textile finishing, 23:905
EO–PO copolymers
  in flotation deinking, 21:17
EPACT. See Energy Policy Act.
EPCRA. See Environmental Planning and
  Community Right-to-Know Act.
EPDM. See Ethylene–propylene–diene
  monomer.

Epichlorohydrin, 21:490
Epidote
  magnetic intensity, 21:876
Epiglotitis
  vaccines against, 24:728
Epinephrine [51-43-4], 22:936; 24:835
Epi-silicon, 23:1066
Episol dyes, 23:361
Epitaxy
  for pure silicon, 21:1094
Epitheaflagallin, 23:756
Epitheaflavic acid, 23:756
Epoxides
  from dimethylsulfonium methylide,
    23:223
Epoxidized natural rubber, 21:576
Epoxy
  silane coupling agent for, 22:152
2-3(3,4-Epoxycyclohexyl)ethyltrimethoxy-
    silane [3388-04-3]
  adhesion promoter, 21:656
Epoxy grouts, 22:457
Epoxy plasticizers, 24:1034
Epoxy resin
  titanated, 24:327
Eprosartan [133040-01-4], 24:47
Epsirantel, 24:832
Epsyn
  ethylene–propylene polymer, 21:486
EPTC [759-94-4], 22:422
  thiols in, 24:29
EQUAL, 23:558
Equalizers, 22:693
Equations of state, 23:990
  in thermodynamics, 23:1012
Equex S
  surfactant, 23:501
Equilenin [517-09-9], 22:852
Equilibrium constant
  effect of SCFs as solvents, 23:469
Ergosterol [57-87-4], 22:856
Erythorbic acid
  oxygen scavenger for steam, 22:742
Erythritol [149-32-6], 23:95
  as sweeteners, 23:556
ESBR. See Emulsion styrene–butadiene
  rubber.
Esca, 21:90
Escherichia coli
  destruction of in foods, 22:848
  effect of silver, 22:174
  inhibited by sorbates, 22:580

Escherichia freundii
  inhibited by sorbates, 22:580
Essential oils, 23:833
  extraction using SCFs, 23:466
  in natural resins, 21:297
Esterification
  organotin catalysts for, 24:145
  of sugar alcohols, 23:101
  of sugars, 23:66
  titanate catalysis, 24:324
Estradiol [50-28-2], 22:859
Estradiol cypronate, 24:833
Estrane [24749-37-9], 22:854
Estra-1,3,5(10)-triene-17-one [900-83-4], 22:876
Estriol [50-27-1], 22:859
Estrogens, 22:852, 890
  as veterinary drugs, 24:834
Estrone [53-16-7], 22:859, 876
Etching
  compound semiconductor processing, 21:798
Ethane [74-84-0], 24:856
1,2-Ethanedithiol [540-63-6], 24:21
3,3-Ethanedyldimercapto-androst-4-ene-11,17-dione [112743-82-5], 22:885
Ethanesulfonic acid [594-45-6], 23:194
Ethanetellurol [83270-38-6], 23:800
Ethanethiol [75-08-1], 24:21
Ethanol [64-17-5], 22:847
  boiling point of, 23:627
  commercial, specific gravity, 23:625
  dielectric constants of, 22:759
  sensors for, 21:823
  separation from water using CO_2, 23:468
  from solar energy, 22:480
  as solvent, 22:536
  specific gravity, 23:625
  supercritical fluid, 23:454
  supercritical fluid cosolvent, 23:457
Ethanolamine [141-43-5]
  as solvent, 22:540
  water additive for steam prdn, 22:742
Ether
  boiling point of, 23:627
Etherification
  of sugar alcohols, 23:101
  of sugars, 23:64
Ethers
  of poly(vinyl alcohol), 24:990

Ethibond Excel, 23:543
Ethilon, 23:543
Ethinylestradiol [57-63-6], 22:903
Ethoduomeen
  surfactant, 23:526
Ethofat
  polyoxyethylene surfactant, 23:514
Ethomeen
  ethoxylate surfactant, 23:526
Ethomid
  polyoxyethylene surfactant, 23:521
Ethopabate [59-06-3], 24:831
Ethoprop
  thiols in, 24:30
Ethyl acetate [141-78-6]
  for decaffeination of tea, 23:762
  as solvent, 22:542
Ethyl acetoacetate [141-97-9]
  as solvent, 22:542
Ethyl alcohol. See *Ethanol*.
Ethylaminoethanol [110-73-6]
  as solvent, 22:540
Ethylbenzene [100-41-4], 22:957
  as HAP compound, 22:532
  rejection by reverse osmosis membrane, 21:319
  separation by reverse osmosis, 21:323
  as solvent, 22:540
ortho-Ethylbenzene [135-01-3], 22:990
Ethyl bromide [74-96-4]
  as solvent, 22:538
2-Ethylbutyl alcohol [97-95-0]
  as solvent, 22:536
Ethyl *t*-butyl ether, 22:480
2-Ethyl-2-butyl-1,3-propanediol [115-84-4], 21:240
Ethyl *n*-butyl sulfate [5867-95-8], 23:410
Ethyl chloride [75-00-3], 24:855
  as solvent, 22:538
Ethyl chlorosulfate [625-01-4], 23:410
Ethyl chlorosulfite [6378-11-6], 23:411
N-Ethylcyclohexyl amine [5459-93-8], 21:468
Ethyldichlorosilane [1789-58-8], 22:47
S-Ethyl dipropylthiocarbamate [759-94-4], 22:427
Ethyldisilane [7528-37-2], 22:42
Ethylene [74-85-1], 24:852
  analysis by optical spectroscopy, 22:640
  sensors for, 21:823
  silver as catalysis for, 22:168
  supercritical fluid, 23:454

Ethylenebis(stearamide) [110-30-5]
as release agent, 21:210
Ethylenediamine, 21:1088
sodium in, 22:330
Ethylenediaminetetraacetic acid (EDTA)
in pulping, 21:14
water additive for steam prdn, 22:741
Ethylenedibiguanidinium sulfate
silver(III) complex of, 22:186
Ethylene dibromide [106-93-4], 22:425
Ethylene glycol [107-21-1], 23:673
as HAP compound, 22:532
as solvent, 22:546
Ethylene glycol dimercaptoacetate
[123-81-9], 24:7
Ethylene glycol dimercaptopropionate
[22504-50-3], 24:7
Ethylene glycol dimethyl ether [110-71-4]
as solvent, 22:544
Ethylene glycol ether esters
as HAP compound, 22:532
Ethylene glycol ethers
as HAP compound, 22:532
Ethylene glycol monobutyl ether [111-76-
2]
as solvent, 22:544
Ethylene glycol monobutyl ether acetate
[112-07-2]
as solvent, 22:542
Ethylene glycol monoethyl ether [110-80-
5]
as solvent, 22:544
Ethylene glycol monoethyl ether acetate
[111-15-9]
as solvent, 22:542
Ethylene glycol monomethyl ether [109-
86-4]
as solvent, 22:544
Ethylene glycol monomethyl ether acetate
[110-49-6]
as solvent, 22:542
Ethylene glycol monophenyl ether [122-
99-6]
as solvent, 22:544
Ethylene glycol monopropyl ether [2807-
30-9]
as solvent, 22:544
Ethylene glycol monopropyl ether acetate
[20706-25-6]
as solvent, 22:542
Ethylene glycol monosalicylate [87-28-5],
21:615

Ethylene interpolymer
as roofing material, 21:448
Ethylene oxide [75-21-8], 24:862
silver cmpd catalysts for, 22:168, 191
for sterilization, 22:843
for sutures sterilization, 23:553
Ethylene–propylene–diene monomer
(EPDM), 21:485
as roofing material, 21:447
Ethylene–propylene rubber, 21:485
Ethylene sulfate [1072-53-3], 23:410
Ethylene sulfite [3741-38-6], 23:410
Ethylenesulfonic acid, 23:169
Ethylenethiourea (ETU) [96-45-7], 21:468;
22:425
Ethyl ether [60-29-7]
as solvent, 22:548
Ethyl 3-ethoxy propionate [763-69-9]
as solvent, 22:544
Ethyl fluorosulfate [371-69-7], 23:410
2-Ethyl-1,3-hexanediol [94-96-2]
as mosquito repellent, 21:241, 244
2-Ethylhexanol [104-76-7]
as solvent, 22:536
2-Ethylhexyl 3-mercaptopropionate
[50448-95-8], 24:7
2-Ethylhexyl salicylate [118-60-5], 21:615
2-Ethylhexyl thioglycolate [7659-86-1],
24:4, 7
Ethyl hydrogen sulfate [540-82-9], 23:416
(17α)-13-Ethyl-17-hydroxy-18,19-
dinorpregnane-4,15-diene-20-yne-3-
one, 22:903
13-Ethyl-17-hydroxy-18,19-dinorpregn-4-
en-20-yn-3-one, 22:903
Ethylmethylethoxysilane [68414-52-8],
22:38
2-Ethyl-4-methyl-1-pentanol [106-67-2]
as solvent, 22:536
Ethylmethylphosphonic acid, 24:509
Ethylphenylsulfinylacetate [54882-04-1],
23:221
Ethyl propionate [105-37-3]
as solvent, 22:542
Ethyl silicate
silica from, 21:1019
Ethylsilicate 40 [18954-71-7], 22:72, 76
Ethylsilicate 50, 22:76
Ethyl stearate
thin films of, 23:1081
Ethyl thioglycolate [623-51-8], 24:7
m-Ethyltoluene [620-14-4], 22:987

o-Ethyltoluene [611-14-3], 22:987
p-Ethyltoluene [622-96-8], 22:987
Ethyltriacetoxysilane [17689-77-9], 22:72
Ethyltriethoxysilane [78-07-8], 22:72
Ethyl trifluoromethane thioglycolate
    [75-92-9], 24:3
Ethyltrimethoxysilane [5314-55-6], 22:72
    critical surface tension of, 22:148
Ethylvanillin [121-32-4], 24:822
Ethyl vinyl ether [109-92-2], 24:1056
Ethyl violet [2390-59-2], 24:560
Etioline [29271-49-6], 22:864
ETU. See Ethylenethiourea.
Eureka process, 23:739
Euxenite [1317-53-9]
    magnetic intensity, 21:876
    tantalum in, 23:660
Evaporation
    steam heat for, 22:758
Evaporators
    for refrigeration, 21:129
    steam heating for, 22:755
    sulfamic acid cleaner for, 23:129
EWG. See Exempt wholesale generator.
Examide
    ethoxylated alcohol surfactant, 23:508
Excess property formulation
    in thermodynamics, 23:1016
Exempt wholesale generator (EWG),
    21:182
Exhaust systems
    titanium dioxide for, 24:238
Expert systems
    use in technical service, 23:780
Explosives
    analysis by γ-ray spectroscopy, 22:656
    gelled, 24:330
    sodium nitrate for prdn of, 22:392
    tellurium in mfg of, 23:805
    trace analysis of, 24:513
    use of sulfur for, 23:252
Extended Nernst-Planck equation
    reverse osmosis models, 21:308
Extenders, 21:527
Extensional viscosity, 21:364
External lubricant, 21:207
Extraction
    DMSO as solvent for, 23:228
    of tea leaf, 23:761
    in trace analysis, 24:493
    use of SCFs, 23:465
Extraction, liquid–liquid
    simultaneous heat and mass transfer,

    22:198
    use of sulfolane, 23:138
Extraction processes
    solvents for, 22:570
Extractive distillation
    use of sulfolane, 23:138
Extractive metallurgy, 22:242
Extraterrestrial materials
    rhenium in, 21:336
Extrinsic semiconductors, 21:726
Extrusion, 22:229
    PVC use, 24:1040
Extrusion cooking
    of starch, 22:700
Eyeglass frames
    of shape-memory alloys, 21:974
Eyring equation, 22:834

F

Fabrics
    cleaning using SCF, 23:465
Fabric softeners, 23:112
Facsimile machines
    amorphous semiconductors in, 21:750
Factice, 23:287
Factor D
    grown in space, 22:622
Factor of safety
    for centrifuges, 21:846
Falling rod viscometer, 21:402
False teeth
    sulfamic acid cleaner for, 23:129
Fan spray atomizers, 22:673
Fansteel process, 23:661
Faraday's law, 23:1069
Farnesol [106-28-5], 23:857
Farnesol acetone [4602-84-0], 23:870
Fascat, 24:146
Fast Green FCF [2353-45-9], 24:564
Fatigue resistance
    of tire cord, 24:164
Fats
    centrifugal separation of, 21:859
    nir spectroscopy of, 22:641
    soap source, 22:297
    sulfation of, 23:171
Fatty acid esters, 23:138
    as release agents, 21:210
    from sucrose, 23:67
Fatty acid neutralization
    in soap manufacture, 22:310

Fatty acids, 23:138
  aspartame in, 23:559
  sampling standards for, 21:627
  soap salts, 22:300
  in tall oil, 23:616
  thin films of, 23:1081
Fatty acid soaps
  in emulsion polymerization, 22:999
  as release agents, 21:207
Fatty acyl glucamides, 23:521
Fatty amides
  as release agents, 21:210
FBC. See Fluidized-bed combustor.
FDA. See Food and Drug Administration.
FD&C Blue No. 2, 23:550
FD&C Green 2 [5141-20-8], 24:556
Febrile respiratory disease
  vaccines against, 24:733
Federal agencies, 21:149
Federal Energy Regulatory Commission
    (FERC), 21:180
Federal Food and Drug Act of 1906,
    24:839
Federal Insecticide, Fungicide, and
    Rodenticide Act (FIFRA), 21:165
Federal Water Pollution Control Act
    (FWPCA), 21:166
Feed, 23:603
  lignosulfonates for, 23:169, 205
  molasses in, 23:603
  precipitated silica, 21:1025
  soybean and other oilseeds for, 22:612
Feed additives
  2-aminoethanesulfonic acid as, 23:210
  lignosulfonates for, 23:169
  molasses in, 23:603
  precipitated silica, 21:1025
  selenium cmpds as, 21:714
  silica as, 21:999
  sodium chloride as, 22:373
  sodium iodide as, 22:382
  sodium sulfates as, 22:410
  soybean and other oilseeds for, 22:612
  sulfur as, 23:261
Feedstocks
  biomass, 22:480
Fehling's solution
  in sugar analysis, 23:15
Feldspars, 21:980
Feline interferon, 24:837
Fellgett advantage, 22:637
Felt
  as roofing material, 21:439

Fenbendazole [43210-67-9], 24:831
α-Fenchene [471-84-1], 23:834
α-Fenchol [1632-73-1], 23:835
Fenchone [1195-79-5], 23:835
Fenchyl alcohol [512-13-0], 23:853
Fenfluramine [458-24-2], 22:938
Fentanyl [437-38-7], 24:835
Ferberite
  magnetic intensity, 21:876
FERC. See Federal Energy Regulatory
    Commission.
Feret diameter, 22:266
Fergusonite
  tantalum in, 23:660
Fermentation
  dextrose for, 23:592
  effect of sodium chloride, 22:372
  unit operation of tea processing, 23:757
Fermented-egg product
  as deer repellent, 21:257
Ferranti-Shirley viscometer, 21:396
Ferric ammonium citrate
  salt additive, 22:368
Ferric chloride [7705-08-0], 24:855
Ferric titanate [1310-39-0], 24:254
Ferrinatrite, 22:404
Ferrite
  in steel, 22:789
Ferroalloys
  role in steelmaking, 22:779
  sampling standards for, 21:628
  silicon in, 21:1114
Ferrophosphorus
  vanadium as by-product, 24:785
Ferroselenium [1310-32-3], 21:698
Ferrosilicon, 21:1113; 22:63
  strontium in, 22:950
Ferrotantalum, 23:676
Ferrotellurium, 23:788
Ferrous chromite, 21:56
Ferrous cupric sulfate
  for chemical dosimeters, 22:845
Ferrous dititanate [12160-10-0], 24:254
Ferrous metatitanate [12168-52-4], 24:254
Ferrous orthotitanate [12160-20-2], 24:254
Ferrous sulfamate, 23:130
Ferrous sulfate
  for chemical dosimeters, 22:845
Ferrous titanate, 24:241
Ferrovanadium alloy [12604-58-9], 24:785
Fertilizers
  raw material for, 23:127

sodium nitrate as, 22:392
tanks for, 23:644
use of sulfur for, 23:251
use of sulfuric acid, 23:397
vanadium as by-product, 24:783
Ferulic acid [437-98-4], 23:563
FET. See Field-effect transistor.
Fiberboard
sampling standards for, 21:627
Fiberfrax, 21:102
Fiber glass, 21:194
Fiber glass felts
as roofing material, 21:443
Fiber glass-reinforced plastic, 23:392
for tanks and pressure vessels, 23:644
Fiber optics
for optical spectroscopy, 22:635, 644
rubidium compounds in, 21:598
shape-memory alloy connectors, 21:969
Fiber optic waveguides
silica, 22:65
Fiber-reinforced rubber
in tire cord, 24:173
Fibers, 22:524; 23:541
PVA resins as, 24:980
sampling standards for, 21:628
from SCF solutions, 23:472
of silica, 21:992, 1000
silica coatings for, 21:1020
by sol–gel technology, 22:500, 524
triarylmethane dyes for, 24:565
use of sulfur for, 23:251
use of sulfuric acid, 23:397
Fibroin, 23:549
Field-effect transistor (FET), 21:778
use as sensors, 21:822
use of selenium, 21:715
Field-flow fractionation
role in particle size measurement,
22:270
FIFRA. See Federal Insecticide, Fungicide,
and Rodenticide Act.
Figs
sorbates in, 22:583
Filaments
rhenium and rhenium alloys as, 21:341
Filament winding, 21:202
Filiform silver, 22:184
Fillers
particle size measurement for, 22:270
in reinforced plastics, 21:195
in staining process, 22:696

Film growth
of amorphous silicon, 21:752
Films
electrically conducting, 23:1049
heat-sealable, 23:112
of polystyrene, 22:1065
silica coatings for, 21:1020
of tellurium, 23:786
Films, anodic oxide
on tantalum, 23:673
Filtering centrifuges, 21:845
Filter media
textile use, 23:882
Filters
surface acoustic wave, 23:677
synthetic quartz crystals for, 21:1082
Filtration
as centrifugal separation, 21:866
sterilization technique, 22:845
in trace analysis, 24:496
water treatment for steam prdn, 22:745
Finasteride [98319-26-7], 22:908
Finazoline
surfactant, 23:527
Fine-grain sugar, 23:41
Fine powders
particle size enlargement, 22:222
Finish distressing, 22:697
Finishers, 21:887
Firebrick
silylating agents for, 22:148
Fireclay
refractories from, 21:49
Fire extinguishers
for tantalum fires, 23:672
Fire-fighting foam
tanks for, 23:644
Firing
unit operation of tea processing, 23:757
Fischer-Tropsch wax [68649-50-3]
as release agent, 21:210
Fish
lipid removal from, 23:467
Fish oils
centrifugal separation of, 21:859
Fixation wire, 23:551
Flame-emission spectrometry
for trace analysis, 24:498
Flame ionization
for trace and residue analysis, 24:500
Flame retardants, 23:908; 24:1036
based on ammonium sulfamate, 23:130
for rubbers, 21:480

Flashing membranes, *21*:452

Flash memories
 silicon-based semiconductors, *21*:744

Flash point
 of air–vapor mixtures, *23*:629

Flat-panel displays
 amorphous semiconductors in, *21*:750
 silicon-based semiconductors, *21*:744

Flatspotting, *24*:178

Flatting agents
 silica gels as, *21*:1023
 silica gels in paint, *21*:1001

Flavan-3,4-diol [*5023-02-9*]
 as bird repellent, *21*:254

Flavan-3,4-*cis*-diol 4-reductase, *23*:751

Flavanols
 in tea, *23*:747

(2*S*)-Flavanone 3-hydroxylase, *23*:751

Flavomycin [*11015-37-5*], *24*:829

Flavone synthase, *23*:751

Flavonoids
 biosynthesis of, *23*:752
 in tea, *23*:747

Flavonols
 in tea, *23*:747

Flavonol synthase, *23*:751

Flavor compounding
 vanillin in, *24*:818

Flavor enhancer
 sweetener as, *23*:573

Flavoring agents
 reactions with aspartame, *23*:560
 for tea, *23*:760
 thiols in, *24*:26
 vanillin in, *24*:818
 vinegar, *24*:838

Flexicoking, *23*:739

Flexographic inks
 from triarylmethane dyes, *24*:557

Flexricin
 glycerol ester surfactant, *23*:512

Flexsorb SE, *23*:439

Flint, *21*:992, 1028

Flint clays
 refractories from, *21*:49

Flocculating agents
 for centrifugal separation, *21*:862
 for sedimentation, *21*:671
 sulfonic acids as, *23*:209
 thioglycolic acid as, *24*:13
 use in silica prprn, *21*:1019
 use in sodium chloride prdn, *22*:361

Flocculation, *21*:671
 use of SCFs, *23*:465
 water treatment for steam prdn, *22*:745

Flooring
 PVC use, *24*:1040
 SBR use, *22*:1011

Floor mats
 from ground rubber, *21*:35

Floors
 silica coatings for, *21*:1020

Florfenicol [*73231-34-2*], [*76639-94-6*]
 as veterinary drugs, *24*:827

Florisil, *24*:496

Flotation
 in paper recycling, *21*:16
 silicates for, *22*:22
 using SCFs, *23*:474

Flotation agents
 sulfonated matls for, *23*:162

Flours
 soybean and other oilseeds, *22*:613

Flour sulfur, *23*:257

Flow, *21*:347
 inside liquid atomizers, *22*:673

Flow curves, *21*:348

Flowers of sulfur, *23*:257

Flow models, *21*:349

Flue gases
 desulfurization sludges, *21*:864
 desulfurized, *23*:447
 removal of sulfur dioxide, *23*:311
 treatment of, *23*:446
 use of sodium with, *22*:346

Fluff, *22*:778

Fluid coking, *23*:738

Fluid energy mills, *22*:291

Fluidized-bed combustor (FBC)
 burning MSW, *21*:230

Fluidized-bed jet mill, *22*:292

Fluidized beds
 adsorption processes, *23*:435
 use for refractory coating, *21*:104

Fluorad, *23*:492

Fluorescein, *22*:190

Fluorescence, *22*:652
 in trace analysis, *24*:492
 use for microemulsion, *23*:462

Fluorescent brighteners
 stilbene dyes as, *22*:927

Fluoride, *23*:285
 in table salt, *22*:373
 in tea, *23*:751

Fluoride glass
  for optical spectroscopy, 22:635
Fluorinated compounds
  as release agents, 21:207
Fluorination
  of organosilanes, 22:53
  use of tantalum fluoride, 23:676
Fluorine
  diffusion in vitreous silica, 21:1048
  etching of silicon, 21:1089
  refractories for, 21:80
Fluorite
  for optical spectroscopy, 22:635
Fluoroacrylates
  polymerization in SCFs, 23:470
Fluoroapatite, 22:64
Fluorocarbon-based refrigerants, 21:139
Fluorocarbons
  coatings for silica gel, 21:1022
Fluorochlorohydrocarbons
  reaction with thorium, 24:80
Fluoroelastomers, 21:490
Fluoroethers
  fractionation using SCFs, 23:469
9α-Fluorohydro cortisone [127-31-1], 22:883
9α-Fluoro-16α-hydroxydrocortisone [337-02-0], 22:884
Fluorometry, 22:653
Fluoropolymers
  as release agent, 21:210
Fluorosilicic acid, 22:65
Fluorosilicones [63148-56-1], 22:110
Fluorosulfuric acid [7789-21-1], 23:195, 203
5-Fluorouracil [51-21-8], 24:836
Fluoxetine [54910-89-3], 22:944
Flurothyl [333-36-8], 22:934
Fluticasone [90566-53-3], 22:906
Fluxes
  for solders and brazing filler metals, 22:496
Foam additives
  sulfonates as, 23:167
Foamed sheet
  of polystyrene, 22:1065
Foamed silicone rubber, 22:113
Foaming-in-place beads, 22:1061
Foams
  blowing by supercritical fluids, 23:460
  polymer from SCFs, 23:473
  rheological measurements, 21:427

FOIA. See Freedom of Information Act.
Foil
  tin, 24:115
Folic acid
  role in veterinary medicine, 24:828
Follicle-stimulating hormone [9034-38-2], 24:833
Folpet [133-07-3], 23:274
Food additives
  regulatory agencies, 21:176
  silica as, 21:999
  sodium iodide as, 22:382
  sodium silicate as, 22:18
  stannous chloride, 24:125, 130
  sugar alcohols as, 23:108
Food and Drug Administration (FDA), 21:168
Food applications
  of sorbates, 22:581
Food Blue 2 [2650-18-2], 24:559
Food Chemicals Codex
  specification for sulfuric acid, 23:399
Food colorants
  triarylmethane dyes for, 24:565
Food packaging
  organotins for, 24:152
  titanium dioxide for, 24:239
Food preservatives
  p-hydroxybenzoic acid in, 21:620
Food processing
  dextrose for, 23:592
  refrigeration in, 21:128
  role of sterilization, 22:849
  sodium silicates in, 22:18
  steel equipment for, 22:823
  use of reverse osmosis in, 21:326
  use of SCFs, 23:466
  use of sodium chloride, 22:372
  use of steam, 22:737
  use of sulfur dioxide, 23:310
Food products
  regulatory agencies, 21:176
  rheological measurements, 21:427
Foods
  drying of, 22:218
  soybeans and other oilseeds for, 22:606
  sterilization techniques for, 22:848
  sugar-free, 23:108
  sulfonic acid-derived dyes in, 23:206
  trace analysis of, 24:515
  use of hexitols in, 23:107
Foot-and-mouth disease, 24:837

Forensic studies
    particle shape classification in, 22:267
Formaldehyde [50-00-0], 24:855
    as liquid sterilant, 22:845
    silver as catalysis for, 22:168
    in textile finishing, 23:896
    use in silica prprn, 21:1019
Formalin
    as liquid sterilant, 22:845
Formate process
    for prdn of sodium dithionite, 23:317
Formic acid [64-18-6], 24:855
    dextrose degradation product, 23:585
    sensors for, 21:823
    in sodium bromide prdn, 22:378
Formulation
    of bar soaps, 22:317
Formvar, 24:936
Formyl chloride [2565-30-2], 24:854
Forsterite [15118-03-3], 21:55
    as refractory material, 21:57
FOS. See Fructosyloligosaccharides.
Fossil copal, 21:298
Foundry furnaces
    refractories for, 21:83
Fourier-transform spectrometer
    for infrared spectroscopy, 22:637
Fractals
    for sol–gel technology, 22:507
Fractionation
    of oils and fats, 23:465
    in paper recycling, 21:18
    SCFs for, 23:468
Fracture mechanics
    particle size reduction, 22:279
Fragrance
    in soap, 22:319
Framed bar process
    for soap, 22:313
Franckeite [12294-04-1], 24:123
Franklinite
    magnetic intensity, 21:876
Frasch mining, 23:433
Frasch process
    for sulfur, 23:240
Freedom of Information Act (FOIA),
    21:165
Free fatty acid
    as soap additive, 22:318
Free-flow agents
    for sodium chloride, 22:368
Free radicals
    microwave spectroscopy of, 22:636

Free-vibration instruments, 21:412
Freight bill, 24:532
Frenkel's theory
    role in sol–gel technology, 22:519
Frequency-modulation spectroscopy,
    22:658
Friction materials
    from ground rubber, 21:35
Friedel-Crafts acylation
    stannic chloride as catalyst, 24:126
Friedel-Crafts reaction
    for styrene, 22:966
    for sulfoxide synthesis, 23:225
Fringed micelle model
    for starch, 22:701
Frings acetator, 24:845
Frit
    as refractory coatings, 21:102
Froude number, 21:674
Frozen rubber, 21:570
α-D-Fructofuranose [10489-79-9], 23:594
β-D-Fructofuranose [470-23-5], 23:594
β-D-Fructofuranosyl-α-D-glucopyranoside
    [57-50-1], 23:1
Fructooligosaccharide, 23:557
α-D-Fructopyranose [10489-81-3], 23:594
β-D-Fructopyranose [7660-25-5], 23:594
Fructose [57-48-7, 30237-26-4], 23:87, 557
    compared to sucrose, 23:4
D-Fructose [57-48-7], 23:87
Fructosyloligosaccharides (FOS), 23:8
Fruit juices
    purification by centrifuge, 21:857
Fruits, 23:96
    dextrose in, 23:583
    sorbates in, 22:583
Fruit sugar, 23:41, 87, 594
Ftir emission spectroscopy, 22:642
Fuchsine [3248-91-7], 24:561
Fuchsonimine hydrochloride [84215-84-9],
    24:553
Fuel additives, 23:209
    titanates as, 24:293
Fuel ash
    refractories for, 21:79
Fuel oils
    zirconium salts of sulfonic acids for,
    23:212
Fuels
    hydrothermal oxidation of, 23:470
    from scrap rubber, 21:24
Fuels, liquid
    thorium as catalyst, 24:71

Fugacity, *23*:1008
  role in modeling phase behavior, *23*:462
Fulgurite, *21*:992, 1033
Fullerene nanosphere
  as release agent, *21*:212
Fullerenes
  sulfonation of, *23*:161
Fulminating silver, *22*:183
Fumed silica, *21*:1026; *22*:63
  silicon tetrachloride in prdn of, *22*:36
Fumes, *22*:296
Fuming sulfuric acid [*8014-95-7*], *23*:363
  by-product of sulfamic acid, *23*:127
Fundamental excess property relation
  of thermodynamics, *23*:1015
Fundamental property relation
  in thermodynamics, *23*:996
Fundamental research
  in industry, *21*:275
Fundamental residual property relation,
  23:1011
Fungi
  centrifugal separation of, *21*:860
  sugar alcohols in, *23*:96
  veterinary drugs against, *24*:829
Fungicides
  DMSO as solvent for, *23*:229
  in soil, *22*:420
Fungistat
  sorbates as, *22*:583
Fur
  triarylmethane dyes for, *24*:565
Furfural [*98-01-1*]
  as solvent, *22*:548
Furfuryl alcohol [*98-00-0*]
  as solvent, *22*:536
Furnace applications
  of refractory fibers, *21*:125
Furnaces
  silica in, *21*:1000
  steel for, *22*:822
  of vitreous silica, *21*:1065
Furnaces, electric
  use in steelmaking, *22*:768
Furnaces, induction
  use in steelmaking, *22*:770
Fusarium episphaeria
  inhibited by sorbates, *22*:579
Fusarium moniliforme
  inhibited by sorbates, *22*:579
Fusarium oxysporum
  inhibited by sorbates, *22*:579

Fusarium roseum
  inhibited by sorbates, *22*:579
Fusarium rubrum
  inhibited by sorbates, *22*:579
Fusarium solani
  inhibited by sorbates, *22*:579
Fusarium tricinctum
  inhibited by sorbates, *22*:579
Fused quartz, *21*:1037
Fused silica, *21*:1027, 1037
Fuses
  shape-memory alloys for, *21*:971
Fusidic acid [*6990-06-3*], *22*:869
Fusion, *22*:229
  vanadium in reactors, *24*:795
FWPCA. See *Federal Water Pollution
  Control Act.*

G

GABA. See *Gamma-aminobutyric acid.*
GABA antagonist
  sulfonic acid derivative as, *23*:211
Gaffix, *24*:1094
Gafquat, *24*:1094
4-*O*-η-D-Galactopyranosyl-D-glucitol, *23*:99
Galactose [*26566-61-0*]
  compared to sucrose, *23*:4
Galai particle size analyzer, *22*:272
Galaxy
  analysis by γ-ray spectroscopy, *22*:656
Galena [*12179-39-4*]
  sulfide ore, *23*:244
Galileo, *23*:1036
Gallium
  in shape-memory alloys, *21*:964
  vitreous silica impurity, *21*:1036
Gallium antimonide [*12064-03-8*], *21*:767
Gallium arsenide [*1303-00-0*], *21*:767, 770;
  22:473
  semiconductors for sensors, *21*:820
  thin films of, *23*:1061
Gallium phosphide [*12063-98-8*], *21*:767
Gamma-aminobutyric acid (GABA)
  picrotoxin as antagonist of, *22*:933
Gamma–phi approach, *23*:1018
Gamma-ray counters
  sodium iodide detectors for, *22*:382
Gamma-ray detector
  metal tellurides as, *23*:805

Gamma-rays
  for sterilization, *22*:845
  use in optical spectroscopy, *22*:656
Ganex, *24*:1094
Ganister, *21*:53
Gantrez AN, *24*:1064
Garlic oil
  sulfoxides in, *23*:217
Garnet
  magnetic intensity, *21*:876
Gas antisolvent recrystallizations
  SCF technique, *23*:472
Gas centrifuge
  separation of uranium isotopes, *21*:871
Gas chromatography
  SCF in sample prprns, *23*:466
  for trace and residue analysis, *24*:500
  use of vitreous silica components,
    21:1065
Gas cleaning
  in sulfuric acid manufacture, *23*:392
Gases
  analysis by multiphoton ionization,
    22:657
  sampling of, *21*:632
Gasifiers
  biomass, *22*:480
Gas–liquid chromatography
  separation of sugar alcohols, *23*:104
Gas, natural. See *Natural gas*.
Gasoline
  nir spectroscopy of, *22*:641
  specific gravity, *23*:625
Gas proportional counters
  use in x-ray spectroscopy, *22*:654
Gas separations
  by polymeric membranes, *23*:460
Gas sweetening, *23*:436
Gas treating
  use of sulfolane, *23*:139
Gas-turbine engines
  refractory coatings for, *21*:105
Gas turbines
  refractory coatings for, *21*:111
  steel for, *22*:823
Gauges
  for vacuum systems, *24*:771
GE-124, *21*:1034
GE-214, *21*:1034
Gelatin, *21*:1020
  sugar alcohols in, *23*:109
Gelatinization
  of starch, *22*:700

Gelation
  of alumina sols, *22*:521
  fractal model, *22*:508
Gelation point, *21*:1021
Gel coats, *21*:196
Gel grouts, *22*:456
GELSIL, *21*:1034
Gel, silica. See *Silica gels*.
Gemini
  vitreous silica windows, *21*:1068
General Conference on Weights and
    Measures, *24*:627
Generators
  steel for, *22*:826
Genetic engineering
  of silkworms and spiders, *22*:161
  in vaccine development, *24*:740
Genistein [446-72-0]
  in soybeans, *22*:598
Gentamycin [1403-66-3], *24*:828
Gentiobiose [554-91-6]
  dextrose degradation product, *23*:585
Gentiopicroside, *24*:500
Geo/Chem AC-400, *22*:455
Geochemistry
  analysis using x-ray fluorescence,
    22:655
Geological materials
  rheological measurements, *21*:427
Geotextiles
  testing of, *23*:947
Geothermal energy
  thermoelectric power sources for,
    23:1036
Geothermal heat pumps (GHP), *21*:233
Geothermal power, *21*:232
Geotrichum candidum
  inhibited by sorbates, *22*:579
Geotrichum sp.
  inhibited by sorbates, *22*:579
GEP. See *Good engineering practice*.
Geranial [141-27-5], *23*:835
Geraniol [106-24-1], *23*:835
Geranonitrile [31983-27-4], *23*:863
Geranyl acetate [105-37-3], *23*:855
Geranyl acetone [396-70-1], *23*:870
Germanium, *21*:107
  thorium oxides, *24*:76
Germanium oxide
  substrate for self-assembled monolayer,
    23:1089
Germanium silicide
  as thermoelectric material, *23*:1035

Germanium tetrachloride [10038-98-9], 24:854
German measles
  vaccines against, 24:728
Germicides, 22:848
Geropon
  dialkyl sulfosuccinate surfactant, 23:498
Gestodene [60282-87-3], 22:903
Gestrogens, 22:891
GHP. See Geothermal heat pumps.
Gibberellic acid [77-06-5], 23:872
Gibbs adsorption isotherm, 23:487
Gibbs-Duhem equation, 23:1004
Gibbs energy
  definition of, 23:996
Gibbsite [14762-49-3], 21:53
Gibbs-Kelvin equation, 22:512
Gibbs's phase rule, 23:1024
Gibbs's theorem, 23:1007
Gilsonite, 21:299
Ginger
  extraction using SCFs, 23:466
Glahard, 24:146
Glance pitch, 21:299
Glass
  bioactive sol-gel materials, 22:526
  frosting with silicon esters, 22:79
  quartz sand for, 21:999
  refractory coatings for, 21:113
  sampling standards for, 21:628
  silicate, 22:5
  silylating agent on, 22:149
  sodium nitrate in prdn, 22:392
  sodium sulfate in prdn, 22:410
  soluble, 22:1
  strontium carbonate in, 22:953
  substrate for self-assembled monolayer, 23:1089
  surface treatment of, 24:126
  tellurium as colorants for, 23:804
  thin films of, 23:1064
  titanium surface coatings, 24:322
  use in optical spectroscopy, 22:635, 644
  use of selenium in, 21:711
  vitreous silica as, 21:1041
Glass capillary viscometer, 21:376
Glass ceramics
  silica for, 21:999
Glass-epoxy composites
  tool materials for, 24:436
Glass fibers, 23:113; 24:323
  silylating agents for, 22:152

by sol-gel technology, 22:500
thermocouple insulation, 23:823
in tire cord, 24:163
Glass frits
  strontium carbonate in, 22:953
Glass, photochromic
  silver cmpds for, 22:192
  silver in, 22:175
Glass-reinforced plastic, 21:194
Glass-reinforced styrene polymers, 22:1028
Glass, soda-lime
  critical surface tension of, 22:148
Glass-transition temperature
  effect of supercritical fluid on, 23:460
Glauberite, 22:404
Glauber's salt, 22:387, 403
Glazes, 22:696
Glazing
  sealant use, 21:665
Glazing interlayer, 24:932
GLC. See Ground-level concentration.
Gliocladium roseum
  inhibited by sorbates, 22:579
Global competition, 21:265
Global warming potential
  of refrigerants, 21:133
Globar
  for optical spectroscopy, 22:635
Glow bar, 23:1044
GLP. See Good laboratory practice.
Glucitol [26566-34-7], 23:95
D-Glucitol [50-70-4], 23:95
D,L-Glucitol [60660-56-2], 23:95
L-Glucitol [6706-59-8], 23:95
Glucoamylase
  for dextrose prdn, 23:586
Glucocorticoids, 22:852
α-D-Glucopyranosyl-4-acetamido-4-deoxy-β-D-glucoheptulopyranoside, 23:72
α-D-Glucopyranosyl-3,4-epoxy-β-D-lyxohexulofuranoside, 23:71
6-O-(α-D-Glucopyranosyl)-β-D-fructopyranose [7158-70-5], 23:75
6-O-(α-D-Glucopyranosyl)-D-fructose, 23:74
4-O-α-D-Glucopyranosyl-D-glucitol, 23:99
6-O-(α-D-Glucopyranosyl)-D-mannitol, 23:74
6-O-(α-D-Glucopyranosyl)-D-sorbitol, 23:74
Glucose [50-99-7]
  compared to sucrose, 23:4

sensors for, *21*:823
in supercritical water, *23*:470
sweetener effect on, *23*:557
α-D-Glucose [*26655-34-5*], *23*:583
β-D-Glucose [*28905-12-6*], *23*:583
Glucose isomerase
for prdn of HFS, *23*:595
Glucose oxidase
in sensors, *21*:823
Glucose syrup, *23*:25, 597
Glucosyl fructose, *23*:569
Glucosyl mannitol [*20942-99-8*]
as sweeteners, *23*:556
Glucosyl sorbitol [*534-73-6*]
as sweeteners, *23*:556
Glues
as paper contaminant, *21*:11
sorbitol in, *23*:112
Glutamic acid
in soybeans and other oilseeds, *22*:596
Glutamic acid–glutamine
in silk, *22*:156
Glutamine
in sugar beets, *23*:54
Glutaraldehyde
as liquid sterilant, *22*:845
Glutathione [*70-18-8*], *24*:20
Glutathione peroxidase, *21*:714
Gluten
centrifugal separation of, *21*:860
effect of sodium chloride, *22*:372
Glycerides
fractionation using SCFs, *23*:466
Glycerin
sampling standards for, *21*:628
Glycerol [*56-81-5*]
salt additive, *22*:368
in soap, *22*:299, 318
Glycerol esters, *23*:512
Glyceryl monothioglycolate [*30618-84-9*],
24:8
Glycetein
in soybeans, *22*:598
3-Glycidoxypropyltrimethoxysilane [*2530-
83-8*]
adhesion promoter, *21*:656
critical surface tension of, *22*:148
silane coupling agent, *22*:150
Glycine
in silk, *22*:156
in soybeans and other oilseeds, *22*:596
strychnine as antagonist of, *22*:933

Glycinin
in soybeans, *22*:595
Glycogen
dextrose from, *23*:583
Glycols
titanates of, *24*:289
Glycomer
as sutures, *23*:542
Glycomer-631, *23*:548
Glycomul SOC
surfactant, *23*:515
Glycomul TO
surfactant, *23*:515
Glycomul TS
surfactant, *23*:515
Glycosperse
sorbitan surfactant, *23*:516
Glycyrrhizic acid [*1405-86-3*]
as sweetener, *23*:568
Glycyrrhizin
as sweetener, *23*:568
Glyphosate [*1071-83-6*], *22*:422
Glysergic acid [*84215-86-1*]
potential sweetener, *23*:573
Goethite
magnetic intensity, *21*:876
Goiter
role of the thyroid, *24*:91
Golay pneumatic cell
for optical spectroscopy, *22*:635
Gold, *21*:107
emissivities of, *23*:827
separation by reverse osmosis, *21*:322
silver as by-product, *22*:171
single dc-diode sputtering of, *23*:1053
substrate for self-assembled monolayer,
23:1089
tellurium in ores, *23*:782
vitreous silica impurity, *21*:1036
Gold alloys
as brazing filler metals, *22*:495
Goldbeater, *23*:545
Gold–cadmium alloys
as shape-memory alloys, *21*:964
Golden syrup, *23*:25
Gold–silver alloys, *22*:176
Gold thiosulfate
complexes of, *24*:65
Golf clubs, *21*:1000
Gonadotrophins
as veterinary drugs, *24*:833
Gonane [*4732-76-7*], *22*:854

Gonorrhea
    developing vaccines for, 24:735
Good engineering practice (GEP), 21:166
Good laboratory practice (GLP), 21:166
Goodrich Disk Fatigue test
    adhesion in tire cord, 24:181
Gore-Tex
    suture, 23:543
Gossypol [303-45-7], 22:609
    in cottonseed, 22:598
Government policies
    and research/technology management,
        21:268
Government Printing Office, 21:166
Government regulations, 21:153
G-protein, 23:578
Grade Quantum 5000, 24:432
Grahamite, 21:299
Grains
    dehulling of, 21:1025
    transportation of, 24:524
Granular material
    rheological measurements, 21:427
Granulated sugar, 23:41
Granulation, 22:279
Granulators, 22:294
Grapes, 23:96
Graphite [7782-42-5], 21:56; 22:176
    for metal freezing point cells, 23:814
    refractory coatings for, 21:105
    as release agent, 21:210
Graphite electrodes
    sampling standards for, 21:627
Graphite tube furnace
    use in optical spectroscopy, 22:645
Grasses
    mannitol in, 23:97
Gratings, holographic
    use in optical spectroscopy, 22:643
Graves' disease, 24:91
Gravitational sedimentation
    role in particle size measurement,
        22:268
Gravity sedimentation, 21:902
Gray cast iron, 23:392
Green acids, 23:497
Greenhouse agriculture, 23:981
Greenhouse effect, 23:453
Green S [3087-16-9], 24:564
Green tea, 23:753
Grignard reagents
    for silanes, 22:60
    for sulfoxide synthesis, 23:225

Grindability, 22:281
Grinding wheels
    sulfur impregnated, 23:263
    tool materials for, 24:437
Griseofulvin [126-07-8], 24:829
Grits
    soybean and other oilseeds, 22:613
Ground-level concentration (GLC), 21:166
Ground rubber, 21:32
Ground-source heat pumps, 21:233
Grouts, 22:452
Growth promoters
    as veterinary drugs, 24:829
GR-S, 22:995
GRS process
    sodium catalyst for, 22:345
Guaiac [9000-29-7], 21:298
Guaiacol [90-05-1]
    vanillin from, 24:813
Guaiol [489-86-1], 23:872
Guaiyl acetate [134-28-1], 23:872
Guanidines
    vulcanization accelerator, 21:463
Guanidine sulfamate [51528-20-2], 23:131
Guggenheim process, 22:385
Guinea Green B [4680-78-8], 24:564
Guinier radius
    use in sol–gel models, 22:508
Gum benzoin [9000-05-9], 21:298
Gum elemi [9000-75-3], 21:297
Gummel number, 21:739
Gum rosin process, 21:292
Gunmetal
    corrosion rates in acid, 23:122
Gymnemic acid [122168-40-5], 23:577
Gypsum
    by-product of titanium dioxide, 24:246
    desulfurization by-product, 23:447
    source of sulfur, 23:246
Gypsum board
    water resistance for, 22:47
Gyratory crushers, 22:285

H

Haake Rotovisco, 21:396
Haematite
    magnetic intensity, 21:876
Hafnium
    in shape-memory alloys, 21:970
Hafnium–tantalum alloys, 21:58

Hagen-Poiseuille expression, *21*:375
Hair preparations
   thioglycolic acid in, *24*:12
Halides
   silver complexes of, *22*:185
Halite [*14762-57-7*], *22*:360
Hallco
   glycerol ester surfactant, *23*:512
Hall coefficient
   of tellurium, *23*:785
Halogen lamps
   sodium iodide for, *22*:382
Halogens
   reactions with silanes, *22*:53
   sensors for, *21*:824
Halosilanes, *22*:61
Halothane [*151-67-7*], *24*:835
Halothiophenes, *24*:41
Hammer mills, *22*:290
Hamposyl 0
   surfactant, *23*:493
Hamposyl C
   surfactant, *23*:493
Hamposyl C-30
   surfactant, *23*:493
Hamposyl L
   surfactant, *23*:493
Hamposyl L-30
   surfactant, *23*:493
Hand lay-up, *21*:196
Hanksite, *22*:404
Hansch multiparameter, *24*:94
Hansenula anomala
   inhibited by sorbates, *22*:580
Hansenula saturnus
   inhibited by sorbates, *22*:580
Hansenula subpelliculosa
   inhibited by sorbates, *22*:580
HAPs. See *Hazardous air pollutants.*
Hardcoats
   silicone products, *22*:116
Hardenability
   of steel, *22*:800
Hardeners
   thioglycolic acids as, *24*:13
Hardening
   of steel, *22*:800
Hardgrove index
   particle size reduction, *22*:281
Hardness. See *Brinell hardness; Knoop
      hardness; Mohs' hardness; Rockwell
      hardness; Vickers hardness.*
   of tool materials, *24*:392

Hardwoods
   xylose from, *23*:105
Hargreaves process, *22*:406
Harmony [*79277-27-3*], *24*:48
Hartamide
   diethanolamine surfactant, *23*:520
Harvade
   thiols in, *24*:30
Hastelloy
   for centrifuges, *21*:845
Hastelloy C-276, *23*:389
Hazard evaluation, *24*:485
Hazardous air pollutants (HAPs), *21*:166;
      22:532
   in power generation, *21*:189
Hazardous Materials Table, *24*:542
Hazardous waste
   transportation of, *24*:543
Hazardous waste analysis
   analysis using x-ray fluorescence,
      22:655
Hazardous waste regulation, *21*:159
Hazelnuts, *23*:96
HBT. See *Heterojunction bipolar
      transistor.*
HDI. See *1,6-Hexamethylene diisocyanate.*
HDK, *21*:1008
HDR. See *Hot dry rock.*
HDR energy, *21*:232
Hearing aids
   silver cmpds for batteries, *22*:190
Heartworm
   veterinary agents against, *24*:831
Heat capacity
   definition of, *23*:999
Heat engine
   in thermodynamics, *23*:986
Heaters, radiant
   silica in, *21*:1000
Heat exchangers
   brazing filler metals for, *22*:490
   sodium, *22*:330
   sulfamic acid cleaner for, *23*:129
   tantalum in, *23*:671
Heating elements
   rhenium and rhenium alloys as, *21*:341
Heat reaction, *22*:229
Heat recovery steam generator (HRSG),
      22:733, 754
Heat regenerators
   refractories for, *21*:83
Heat reservoir
   in thermodynamics, *23*:986

Heat shields
    tantalum in, 23:671
Heat transfer
    with simultaneous mass transfer,
        22:196
    for tanks, 23:647
Heat-transfer fluids
    organotin catalysts for, 24:145
Heat-transfer media
    silicon esters as, 22:79
    sodium as, 22:345
Heavy metal
    extracting using SCF, 23:465
Heavy oil, 23:717, 718
Hecogenin [467-55-0], 22:862
Hectorite, 22:26
Hedge parsley
    mannitol in, 23:97
Heliostats, 22:471
Helium
    diffusion in vitreous silica, 21:1048
    removal from natural gas, 23:139
Helmholtz energy
    definition of, 23:996
Helminthosporium sp.
    inhibited by sorbates, 22:579
Helvolic acid [29400-42-8], 22:869
Hemagglutinins
    in soybeans and other oilseeds, 22:596
Hemimorphite
    silylating agents for, 22:147
Hemiterpene, 23:833
Hennecke, 24:710
Henry's law, 24:258
Hepatitis B, 24:732
Heptachlor [76-44-8], 22:421
    regulatory level, 21:160
1,3,5,7,9,11,13-Heptaethylcyclohepta-
        siloxane [17909-36-3], 22:103
Heptane [142-82-5]
    as solvent, 22:536
1-Heptanethiol [1639-09-4], 24:21
Herbicides, 23:113, 466
    ammonium sulfamate in, 23:130
    DMSO as solvent for, 23:229
    in soil, 22:420
    trace analysis of, 24:495, 503, 510
Hercules viscometer, 21:397
Hercynite, 21:56
Hernandulcin [108944-70-3]
    potential sweetener, 23:573
Herpes simplex, 23:211
    vaccines for, 24:735, 737

Herschel-Bulkley model, 21:350
Hertz equation, 21:409
Herzenbergite [14752-27-3], 24:123
Hessite [12002-98-1]
    tellurium in, 23:782
Heteroarylmethane dyes, 24:567
Heterojunction bipolar transistor (HBT),
        21:782
Heterojunction FET, 21:780
Heterosporium terrestre
    inhibited by sorbates, 22:579
Hevea brasiliensis, 21:562
Heveacrumb process, 21:567
Hexaacetoxydititanoxane [4861-18-1],
        24:298
Hexachlorobenzene [118-74-1], 22:425
    regulatory level, 21:160
Hexachlorobutadiene [87-68-3]
    regulatory level, 21:160
Hexachlorodisilane [13465-77-5], 22:43
Hexachloroethane [67-72-1], 24:832
    regulatory level, 21:160
Hexachlorophene, 22:848
Hexadecamethylcyclooctasiloxane
        [556-68-3], 22:103
Hexadecamethylheptasiloxane [541-01-5],
        22:103
1-Hexadecanethiol [2917-25-2], 24:21
n-Hexadecyl hydrogen sulfate [143-02-2],
        23:410
trans,trans-2,4-Hexadienoic acid [110-44-
        1], 22:571
Hexaethoxydisiloxane [2157-42-8], 22:72
Hexafluorodisilane [13830-68-7], 22:43
Hexafluoroisopropyl alcohol [920-66-1],
        24:277
Hexahydro-1-((2-methylcyclohexyl)-
        carbonyl)-1*H*-azepine [52736-62-6]
    as repellant, 21:249
Hexamethoxymethylmelamine [3089-11-
        0], 21:479
Hexamethylcyclotrisiloxane [541-05-9],
        22:103
Hexamethyldisilane [1450-14-2], 22:145
Hexamethyldisilazane [999-97-3]
    silylating agent, 22:144
Hexamethyldisiloxane [107-46-0], 22:103
Hexamethylditin [661-69-8], 24:148
Hexamethylene-1,6-bisthiosulfate,
        disodium salt [5719-73-3]
    cross-linking agent, 21:471

Hexamethylene diamine carbamate
  [143-06-0]
  cross-linking agent, 21:470
Hexamethylene diisocyanate (HDI)
  [822-06-0], 24:696, 707
Hexamethylenetetramine [100-97-0],
  21:468
Hexane [110-54-3]
  for processing soybeans and other
    oilseeds, 22:601
  as solvent, 22:536
N-Hexane [110-54-3]
  as HAP compound, 22:532
n-Hexane–2-propanol–water system,
  21:933
Hexanesulfonic acid [13595-73-8], 23:194
1-Hexanethiol [111-31-9], 24:21
1,2,6-Hexanetriol trithioglycolate
  [19759-80-9], 24:13
1-Hexanol [111-27-3]
  as solvent, 22:536
1,4,8,11,15,18-Hexaoctyl-22,25-bis-
  (carboxypropyl)-phthalocyanine,
  23:1082
Hexyl alcohol [111-27-3]
  as solvent, 22:536
Hexylene glycol [107-41-5]
  as solvent, 22:546
Hi-Activity process, 23:442
Hides
  dehairing with sodium sulfide, 22:417
High amylose corn
  starch from, 22:699
High compression roll mills, 22:286
High electron mobility transistors, 21:780
High fructose corn syrup, 22:699; 23:25,
  557, 582
  compared to cane sugar, 23:39
High impact polystyrene (HIPS), 22:1016
High performance thin-layer
  chromatography, 24:499
High performance tires
  rayon in, 24:163
High pressure liquid chromatography
  for trace and residue analysis, 24:501
High temperature alloys
  brazing of, 22:484
  as refractory coatings, 21:94
  vanadium in, 24:794
High temperature resistance
  of refractory fibers, 21:125
High test molasses, 23:42

Hinder, 21:258
HIPS. See High impact polystyrene.
HIPS materials, 22:1020
Hi-Sil 190, 21:1008
Histidine
  in silk, 22:156
  in soybeans and other oilseeds, 22:596
HIV
  developing vaccines for, 24:735
HIV-1 reverse transcriptase
  grown in space, 22:622
Hodag
  glycerol ester surfactant, 23:512
  polyoxyethylene surfactant, 23:514
  sorbitan surfactant, 23:516
Hodag Amine C
  surfactant, 23:527
Hodag DGL
  fatty acid surfactant, 23:518
Hodag DGO
  fatty acid surfactant, 23:518
Hodag DGS
  fatty acid surfactant, 23:518
Hodag PMS
  fatty acid surfactant, 23:518
Hodag SML
  surfactant, 23:515
Hodag SMO
  surfactant, 23:515
Hodag SMP
  surfactant, 23:515
Hodag STO
  surfactant, 23:515
Hodag STS
  surfactant, 23:515
Hoeppler viscometer, 21:400
Hog fat
  in soap manufacture, 22:302
H-Oil process, 23:739
Hollow-cathode lamps
  use in optical spectroscopy, 22:646
Hollow-cone sprays, 22:676
Hollow-fiber membranes
  in ultrafiltration, 24:621
Hollow fibers, 24:333
Hollow metal fibers
  for optical spectroscopy, 22:635
Homer Mammalok, 21:974
Home scrap, 22:778
Homomenthyl salicylate [118-56-9], 21:614
Homosalate [118-56-9], 21:615
Homosil, 21:1034, 1049

Honey
    dextrose in, 23:583
Hormonal steroids, 22:852
Hormones
    as veterinary drugs, 24:833
Hornblende
    magnetic intensity, 21:876
Hoses
    precipitated silica in, 21:1025
Hostapur, 23:494
Hot dry rock (HDR)
    geothermal resource, 21:232
Hot extrusion process
    for soap, 22:317
Hot GR-S, 22:1004
Hot-water process, 23:732
HRSG. See Heat recovery steam generator.
HSH syrups, 23:100
H steels, 22:801
Hubble telescope, 21:1068
Hudson-Bertrand rules, 23:102
Human growth hormone
    detection of, 24:506
Human α-thrombin
    grown in space, 22:622
Humectants
    in adhesives, 24:971
    for sodium chloride, 22:368
    sorbitol solutions as, 23:108
Humicola fusco-atra
    inhibited by sorbates, 22:579
Humidifiers
    sulfamic acid cleaner for, 23:129
Humidity, 22:199
    sensors for, 21:824
Hutchinsonite [12198-34-4], 23:952
H values
    of steel, 22:801
Hydrated tantalum oxide [75397-94-3],
    23:665
Hydraulic fluids
    silicon esters as, 22:79
Hydrazine
    in steam, 22:737
    water additive of steam prdn, 22:740
Hydridorhenium pentacarbonyl [16457-
    30-0], 21:343
Hydrobromic acid [10035-10-6]
    sodium bromide from, 22:378
Hydrocarbons, 21:825; 23:466
    as refrigerant, 21:131, 134
    removal of, 23:436

    as silica coating, 21:1020
    sodium dispersions in, 22:328
    tanks for, 23:625, 631
    thorium as catalyst, 24:71
Hydrocarbon solvents, 22:535
Hydrochloric acid [7647-01-0], 21:695
    metal corrosion rates, 23:122
    from sodium chloride, 22:374
    in sodium chloride prdn, 22:361
    in steam, 22:732
Hydrochlorofluorocarbons
    as refrigerant, 21:131, 134
Hydrocyclones, 21:902
    for size separation, 21:911
Hydrodesulfurization, 23:281
Hydrodynamic process
    for sulfur, 23:242
Hydroelectric power, 21:233
Hydrofluorene [34069-54-0]
    potential sweetener, 23:573
Hydrofluoric acid, 21:1041
    etching of silicon, 21:1089
    removal by reverse osmosis, 21:322
    for silica prprn, 21:1019
    use of sulfuric acid, 23:397
Hydrofluorocarbons
    as refrigerant, 21:131, 134
Hydrogels, 21:996, 1020; 22:25
    from sucrose, 23:7
Hydrogen [1333-74-0]
    diffusion in vitreous silica, 21:1048
    prdn by steam reforming, 22:757
    refractories for, 21:79
    sensors for, 21:823
    as steam component, 22:721
    from titanium hydrides, 24:227
    use in optical spectroscopy, 22:645
Hydrogenated castor oil [8001-78-3]
    as release agent, 21:210
Hydrogenated starch hydrolysates, 23:99
Hydrogenation, 21:343
    catalytic for sugar alcohols, 23:105
    thorium as catalyst, 24:71
Hydrogen bonding
    effect on SCF phases, 23:463
Hydrogen bromide [10035-10-6], 24:855
Hydrogen chloride [7647-01-0], 24:852
Hydrogen energy
    role of thermoelectric energy, 23:1037
Hydrogen fluoride [7664-39-3], 24:854
    tetrafluorosilane as by-product, 22:64
Hydrogen iodide [10034-85-2], 24:855

Hydrogen-ion activity
  control agents for steam, 22:737
  sensors for, 21:823
Hydrogen peroxide, 21:695; 22:846
  in paper recycling, 21:18
  in pulping, 21:14
  silicates with, 22:24
  water additive of steam prdn, 22:740
Hydrogen polysulfides, 23:284
Hydrogen selenide [7783-07-5], 21:688, 699
  removal from fossil fuels, 23:139
Hydrogen sulfide [7783-06-4], 23:275; 24:854
  sulfuric acid from, 23:377
  sulfur removal and recovery from, 23:433
Hydrogen telluride [7783-09-7], 23:793
Hydrol, 23:589
Hydrolysis
  of fats and oils, 22:310
  role in sol–gel technology, 22:505
  of starch, 22:704
Hydrolysis resistance
  in finished textiles, 23:897
Hydrolyzed cereal solids, 23:600
Hydrometallurgy
  for sulfur, 23:245
Hydrometer
  in sugar density measurement, 23:15
Hydron Blue G
  sulfur dye, 23:342
Hydron Blue R
  sulfur dye, 23:342
Hydrophobic silica [7631-86-9]
  as release agent, 21:212
Hydroponics
  super absorbent polymers, 22:460
Hydropower, 21:234
Hydroquinone [123-31-9], 24:79, 876
  oxygen scavenger for steam, 22:742
Hydroseeding
  super absorbent polymers, 22:460
Hydrosilylation, 22:54, 56, 86
Hydrostatic pressure
  role in tank design, 23:642
Hydrothermal oxidation
  use of supercritical water, 23:470
Hydrothermal processing
  of synthetic quartz crystals, 21:1079
  use of steam, 22:759
Hydrotropes
  sulfonated aromatics as, 23:159

9α-Hydroxyandrosta-4-ene-3,17-dione [560-62-3], 22:878
o-Hydroxybenzamide [65-45-2], 21:616
2-Hydroxy-1,3-benzenedicarboxylic acid [606-19-2], 21:605
4-Hydroxy-1,3-benzenedicarboxylic acid [636-46-4], 21:605
m-Hydroxybenzoic acid [99-06-9], 21:601
o-Hydroxybenzoic acid [69-72-7], 21:601
p-Hydroxybenzoic acid [99-96-7], 21:601
(m-Hydroxybenzoyl)tropeine [52418-07-2], 21:620
o-Hydroxybenzyl alcohol [90-01-7], 21:621
3α-Hydroxycholanic acid [434-13-9], 22:856
Hydroxycitronellal [107-75-5], 23:835
Hydroxycitronellol [107-74-4], 23:858
3-Hydroxycyclohexanecarboxylic acid [22267-35-2], 21:619
17η-Hydroxy-11η-(4-dimethylamino-phenyl-1)-17α-(prop-1-ynyl)-estra-4,9-diene-3-one, 22:903
20-Hydroxyecdysone [5289-74-7], 22:868
2-Hydroxyethyl acrylate [818-61-1]
  in grouts, 22:455
Hydroxyethylstarch [9005-27-0], 22:713
3-Hydroxy-4-iodobenzoic acid [58123-77-6], 21:619
2-Hydroxyisoflavanone synthase, 23:751
Hydroxyl
  vitreous silica impurity, 21:1036
Hydroxylalkyl starch ethers, 22:710
Hydroxylamine-O-sulfonic acid [2950-43-8], 23:211
5-Hydroxymethylfurfural [67-47-0]
  dextrose degradation product, 23:585
3-Hydroxy-2-nitrobenzoic acid [602-00-6], 21:619
3-Hydroxy-4-nitrobenzoic acid [619-14-7], 21:619
(17α)-17-Hydroxy-19-norpregn-4-en-20-yn-3-one, 22:903
3-Hydroxy-2-oxetanone, 22:846
4-Hydroxyphthalide [13161-32-5], 21:619
α-Hydroxy-O-propiolactone, 22:846
Hydroxypropylstarch [9049-76-7], 22:714
3-Hydroxysulfolane [13031-76-0], 23:137
11-Hydroxyundecane-thiol
  self-assembled monolayers on gold, 23:1106
Hydrozincate
  as rubber chemical, 21:473

Hyonic
  ethoxylated alkylphenol surfactant,
    23:511
Hypalon, 23:394
Hyperlinks, 23:780
Hyphenated instruments
  including infrared detection, 22:640
Hypo, 24:57
Hypochlorites, 22:847
Hypophosphorous acid
  in selenium analysis, 21:707
HYSOLAR program, 23:1037
Hysulf process, 23:445

I

IBP
  thiols in, 24:30
ICC. See *International Chamber of
    Commerce.*
Ice cream, 23:110
  corn syrups in, 23:601
Ideal gas, 23:1021
Ideal solution
  in thermodynamics, 23:1013
Iditol [24557-79-7], 23:95
D-Iditol [23878-23-3], 23:95
L-Iditol [488-45-9], 23:95
IFPEXOL, 23:440
Igepal
  ethoxylated alkylphenol surfactant,
    23:511
Ignition supression agents
  in styrene plastics, 22:1058
Ilmenite [12168-52-4], 24:230, 241
  magnetic intensity, 21:876
  titaniums in, 24:225
Ilmenorutile
  magnetic intensity, 21:876
Image analyzers
  use for size measurement of particles,
    22:266
Imaging agents, 24:493
Imaging arrays
  use in optical spectroscopy, 22:643
Imaging technology
  use for size measurement of particles,
    22:265
Imass Dynastat, 21:418
Imazapyr [81335-77-5], 22:448
Imazaquin [81335-37-7], 22:448

Imazethapyr, 22:441
Imidazolines, 22:447
Imipramine [50-49-7], 22:941
Immedial Black V
  sulfur dye, 23:342
Immedial Bordeaux 3BL
  sulfur dye, 23:342
Immedial Brown GGL
  sulfur dye, 23:342
Immedial Katechu 4RL
  sulfur dye, 23:342
Immedial Maroon B
  sulfur dye, 23:342
Immedial New Blue 5RCF
  sulfur dye, 23:342
Immedial Orange C
  sulfur dye, 23:342
Immedial Pure Blue
  sulfur dye, 23:342
Immedial Yellow Brown GL
  sulfur dye, 23:342
Immiticide, 24:831
Immobile liquids
  binder, 22:224
Immunization, 24:727
Immunoassays, 24:508
  organoselenium compounds in, 21:714
Immunology, 24:740
Immunotherapeutic agents
  as veterinary drugs, 24:837
Impact
  for size reduction, 22:283
Impact crusher, 22:289
Impact mills, 22:289
IMPATTs, 21:777
Imperialine [61825-98-7], 22:865
Incineration, 22:848
Incinerators
  refractories for, 21:84
Inclined-disk agglomerator, 22:232
Inconel, 23:281
  for tantalum reduction reactors, 23:666
Inconel 718
  tool materials for, 24:431
IND. See *Investigational new drug.*
Indalone [532-34-3]
  as mosquito repellent, 21:244
Indentometers, 21:409
Indium
  in solders, 22:485
  vapor pressure of, 23:1046
Indium alloys
  as shape-memory alloys, 21:964

Indium antimonide [1312-41-0], 21:767
Indium arsenide [1303-11-3], 21:767
Indium phosphide [22398-80-7], 21:767
Indium–tin oxide
  as refractory coating, 21:113
  thin films of, 23:1073
Indocarbon CL
  sulfur dye, 23:342
Indolyldiphenylmethane dye, 24:556
Inductively coupled plasma
  for optical spectroscopy, 22:647
Industrial solvents, 22:529
Inert gas
  tank blanketing systems, 23:630
Influenza
  vaccine against, 24:738
Infrared absorption
  of sol–gels, 22:517
Infrared detectors
  metal tellurides as, 23:805
  tellurium as, 23:783
Infrared spectroscopy, 22:636
Infrared Vitreosil, 21:1049
Infrasil, 21:1034
Infrasil fused quartz
  diffusion of ^{22}Na, 21:1047
Inherent viscosity, 21:356
Inhibitors
  for styrene, 22:973
Inhibitory activity
  of sorbates, 22:578
Injection molding
  reinforced plastics, 21:201
Ink jet, 24:329
Ink jet printing
  triarylmethane dyes, 24:565
Inks
  as paper contaminant, 21:11
  selenium as colorant for, 21:712
  solvents for, 22:570
  sulfonic acid-derived dyes in, 23:206
Inorganic, 24:225
Inorganic esters
  of poly(vinyl alcohol), 24:989
Inositol 1,4,5-trisphosphate
  role in sweetness, 23:578
Insecticides, 23:113
  DMSO as solvent for, 23:229
  in soil, 22:420
Insect repellents, 21:236
In situ diagnostics
  in MOCVD, 21:772

In situ processing, 23:730
Insoluble sulfur [9035-99-8]
  cross-linking agent, 21:470
Instantizer agglomerator, 22:236
Instruments, precision
  silica in, 21:1000
Insulating glass
  sealant use, 21:665
Insulation
  PUIR foam, 24:724
  refractories for, 21:47
  in refrigerators, 21:135
  silica in, 21:1002
  for tanks, 23:647
  textile use, 23:882
Insulation, electric
  PVF resins, 24:938
Insulator
  talc, 23:611
Insulator, electric
  vitreous silica as, 21:1055
Insulin response
  for sugar alcohols, 23:107
Integrated circuits, 21:720
  pure silicon for, 21:1102
  role in sensors, 21:820
  vitreous silica in prdn of, 21:1067
Interfacial energy, 23:483
Interferometers
  for infrared spectroscopy, 22:637
Interferons, 24:837
Interlevel wiring, 21:807
Intermediate 300
  surfactant, 23:520
Intermediate 325
  surfactant, 23:520
Intermetallic compounds
  in steel, 22:824
Intermolecular forces
  binder, 22:224
Internal energy
  in thermodynamics, 23:985
International Association for Properties of
    Water and Steam, 22:720
International Association for the
    Properties of Steam, 22:720
International Chamber of Commerce
    (ICC), 21:154
International Institute of Synthetic
    Producers, 21:482
International Natural Rubber
    Specification, 21:483

International Organization for
Standardization (ISO), *21*:76
International Sugar Scale, *23*:13
International System of Units, *24*:628
Internet
use in technical service, *23*:779
Interstate commerce, *24*:533
Interstate Commerce Commission, *24*:527
Interstitial cystitis
DMSO for, *23*:227
Interstitial-free steels, *22*:816
Inthion Brilliant Blue I3G
sulfur dye, *23*:343
Intrastate commerce, *24*:533
Intrinsic semiconductors, *21*:725
Intrinsic viscosity, *21*:357
Inulin, *23*:594
Invert molasses, *23*:603
Invert sugar, *23*:87
sorbitol from, *23*:105
Invert syrup, *23*:25
Investigational New Drug (IND), *21*:171
Iodargyrite, *22*:179
Iodine [*7553-56-2*], *23*:196; *24*:866
in table salt, *22*:373
Iodine complexes
of PVP, *24*:1088
Iodoform [*75-47-8*], *24*:856
Iodophores, *22*:847
Iodosilanes
for microelectronics, *22*:47
Ion chromatography, *24*:504
for water analyses in steam prdn,
22:746
Ion exchange, *22*:745; *24*:516
metal ions and silicates, *22*:10
in steam prdn, *22*:745
Ion-exchange chromatography, *24*:504
for trace and residue analysis, *24*:501
Ion-exchange materials
sampling standards for, *21*:627
Ion-exchange resins
for sugar alcohols, *23*:104
sulfonic acid derivatives for, *23*:169
Ion-exclusion chromatography
to remove sugar from beet molasses,
23:603
Ion implantation
compound semiconductor devices,
21:801
for refractory coatings, *21*:98
Ionizing radiation detectors
sensors for, *21*:820

Ionomeric pendent groups
in PVB resins, *24*:928
trans-α-Ionone [*6901-97-9*], *23*:835
α-Ionone [*127-41-3*], *23*:864
β-Ionone [*14901-07-6*], *23*:835, 864
γ-Ionone [*79-6-5*], *23*:864
Ion plating, *23*:1056
Ion product
of water, *22*:725
Ion-selective electrodes
determination of silver, *22*:187
as sensors, *21*:825
for water analyses in steam prdn,
22:745
Iproniazid [*54-92-2*], *22*:939
Irehidiamine-A [*3614-57-1*], *22*:864
Irganox 1076
rubber antioxidants, *22*:1023
Irgasan 300, *24*:510
Iridium
as refractory coating, *21*:91
Irish Spring, *22*:323
Iron
adhesion in tire cord, *24*:171
in brazing filler metals, *22*:492
complexes with sugar alcohols, *23*:103
corrosion rates in acid, *23*:122
critical surface tension of, *22*:148
emissivities of, *23*:827
ferrosilicon in prdn of, *21*:1114
impurity in selenium, *21*:705
magnetic separation of, *21*:878
role in steel, *22*:765
in silicon and silica alloys, *21*:1106
in steam, *22*:737
in sulfuric acid, *23*:399
sulfur in prdn of, *23*:260
tellurium-chilled, *23*:801
titanium dioxide ore by-product, *24*:242
vanadium in, *24*:794
vitreous silica impurity, *21*:1036
Iron alloys
as shape-memory alloys, *21*:965
use with steam, *22*:761
Iron carbide, *22*:766
role in steelmaking, *22*:767
Iron castings
strontium as inoculant, *22*:950
tellurium for control of chill depth,
23:801
Iron–chromium–nickel alloys
steels from, *22*:817

Iron-constantan
  thermocouples, 23:823
Iron exchange
  hydrated titanium dioxide for, 24:235
Iron–iron carbide
  phase diagram, 22:790
Iron–manganese–silicon
  shape-memory alloys, 21:975
Iron ore
  sampling standards for, 21:627
Iron oxide [1317-61-9]
  sorbent for hydrogen sulfide, 23:434
Iron oxides
  cause of boiler tube scale, 22:738
  sol surfaces, 22:11
  solubility in steam, 22:728
Iron pentacarbonyl [13463-40-6], 24:856
Iron pyrites
  sulfuric acid from, 23:377
Iron sulfide [1317-37-9], 23:434
Iron telluride [12125-63-2], 23:788
Iron titanate brown, 24:248
ISO. See International Organization for
  Standardization.
ISO 14000, 21:154
Isoamyl acetate [123-92-2]
  as solvent, 22:542
Isoamyl salicylate [87-20-7], 21:614
Isoborneol [124-76-5], 23:835
Isobornyl acetate [125-12-2], 23:846
Isobornyl methyl ether [5331-32-8], 23:834
Isobutyl acetate [110-19-0]
  as solvent, 22:542
Isobutyl isobutyrate [97-85-8]
  as solvent, 22:542
Isobutyl salicylate [87-19-4], 21:614
Isobutyltrimethoxysilane [18395-30-7],
  22:72
3-(2-Isocamphyl) cyclohexanol [70955-71-
  4], 23:846
Isocarboxazid [59-63-2], 22:939
Isocyanates
  polyurethanes, 24:703
  from poly(vinyl alcohol), 24:990
3-Isocyanatomethyl-3,5,5-
  trimethylcyclohexyl isocyanate
  [4098-71-9], 24:708
Isocyanide
  thorium complexes, 24:74
Isoelectric point
  role in sol–gel technology, 22:509
Iso E Super [54464-57-2], 23:848

Isoflurane [26675-46-7], 24:836
Isoglucose, 23:594
Isoidide, 23:100
Isoleucine
  in silk, 22:156
  in soybeans and other oilseeds, 22:596
(+)-trans-Isolimonene [5133-87-1], 23:843
Isomalt [64519-82-0], 23:95
  as sweeteners, 23:556
Isomaltitol, 23:74
Isomaltose [499-40-1]
  dextrose degradation product, 23:585
Isomaltulose, 23:74
Isomannide, 23:100
α-Isomethylionone [7779-30-8], 23:865
Isomethylpseudoionone [1117-41-5],
  23:865
Isooctyl 3-mercaptopropionate [30774-
  01-7], 24:7
Isooctyl thioglycolate [25103-09-7], 24:4, 7
Isophorone [78-59-1]
  as HAP compound, 22:532
  as solvent, 22:546
Isophorone diisocyanate [4098-71-9],
  21:1000
Isophytol [505-32-8], 23:857
trans-Isophytol [150-86-7], 23:873
Isopimaric acid [5835-26-7], 21:294
Isoplast, 24:718
Isoprene [78-79-5], 23:139
  in terpene synthesis, 23:836
Isopropanol [67-63-0]
  supercritical fluid, 23:454
Isopropyl acetate [108-21-4]
  as solvent, 22:542
Isopropyl alcohol, 21:1088
Isopropylamine [75-31-0]
  as solvent, 22:540
Isopropylation
  of toluene, 24:357
Isopropyl chlorosulfate [36610-67-0],
  23:411
Isopropyl chlorosulfite [22598-56-7],
  23:411
p-Isopropyl-N,N-dimethylbenzamide
  [6955-06-2]
  as repellent, 21:249
Isopropyl ether [108-20-3]
  as solvent, 22:548
Isopropylmethylphosphonic acid, 24:509
Isopropyl propionate
  as solvent, 22:542

Isopropyl salicylate [607-85-2], 21:614
Isopropyl titanate, 24:238
Isopropyl vinyl ether [926-65-8], 24:1056
Isoquinoline
    methylation of, 23:223
1-Isoquinolone [491-30-5], 24:856
Isosorbide, 23:100
Isosyrup, 23:594
Isovanillin [621-59-0], 23:571
Isoxicam, 24:515
Itabirite
    magnetic intensity, 21:876
Itraconazole [84625-61-6]
    as veterinary drug, 24:829
Ivermectin [70288-86-7], 24:830
Ivomect, 24:832
Ivory, 22:323

J

Jacobsen rearrangement, 23:159
Jalaric acid, 21:300
Japanese National Large Telescope,
    21:1068
Jasmine
    in tea, 23:759
Jaw crushers, 22:285
Jeffrey's rule, 23:94
Jergens Mild, 22:323
Jervine [469-59-0], 22:865
Jet fuel
    centrifugal separation of, 21:859
Jet mills, 22:291
Jet spray, 22:677
JO-1222 [98449-05-9], 22:906
Johnson noise, 23:828
Jominy test, 22:801
Joosten method
    silicate grouting, 22:453
Junction devices
    compound semiconductors in, 21:796
Junction FET, 21:779

K

Kaempferol [520-18-3]
    in tea, 23:749
Kainate, 22:935
Kainic acid [487-79-6], 22:935
Kallidin, 24:506

Kanamycin [59-01-8], 24:828
Kanemite [38785-33-0], 22:5
Kaolin [1332-58-7]
    refractories from, 21:49
    as release agent, 21:210
Kaolinite [1318-74-7], 21:53
Kaolizer, 24:700
Karbate cooler/condensers, 23:391
Kashin-Beck disease, 21:714
Kathon
    thiols in, 24:30
Kauri-Butanol value, 21:298; 22:535
Keatite [17679-64-0], 21:988
Kedem-Katchalsky
    reverse osmosis models, 21:308
Kel-F, 24:32
Kemamine
    surfactant, 23:529
Ken-React KR TTS, 24:330
Kenyaite [12285-95-9], 22:5
Keratin, 24:12
Keratolytic agents
    salicylic acid derivatives, 21:610
Kerosene
    specific gravity, 23:625
Keshan disease, 21:714
Kessco
    glycerol ester surfactant, 23:512
Kessco EGDS
    fatty acid surfactant, 23:518
Kessco EGMS
    fatty acid surfactant, 23:518
Kestner-Johnson dissolver, 22:183
17-Ketal [2398-63-2], 22:876
Ketals
    from sugar alcohols, 23:102
Ketamine hydrochloride [1867-66-9],
    24:835
Ketene-crotonaldehyde route
    to sorbic acid, 22:576
4-Ketoflavan-3-ol [577-85-5]
    as bird repellent, 21:254
Ketoprofen [22071-15-4], 24:502, 832
11-Keto-progesterone [516-15-4], 22:894
Ketoses
    from sugar alcohols, 23:102
Ketotifen [34580-13-7], 24:48
11-Ketotigogenin [4802-74-8], 22:875
Kettle boiling batch process
    for saponification, 22:307
Kevlar
    mechanical properties of, 22:160

Keyboards
  silver switches for, *22*:176
Kilns
  refractories for, *21*:83
Kinematic viscosities, *21*:382
Kinetic measurements
  optical spectroscopies for, *22*:659
Kinetics
  effect of SCFs as solvents, *23*:469
Klystrons, *22*:636
Kneading
  in paper recycling, *21*:17
Knecht method, *24*:556
Knife mills, *22*:294
Knight and Allen
  reducing sugar test, *23*:16
Knives
  refractory coatings for, *21*:113
Knoop hardness
  of borides, *21*:106
  of diamond, *24*:436
  of pure silicon, *21*:1096
  of titanium silicides, *24*:263
  of vitreous silica, *21*:1054
  titanium monophosphide, *24*:264
Kocheshkov redistribution reaction,
    24:133
Kollidon, *24*:1094
Kraft lignin, *23*:168
Kraft pulping process
  source of dimethyl sulfide, *23*:225
  tall oil recovery, *23*:618
Krebs cycle acids
  chromatography of, *22*:145
Krieger-Dougherty equation, *21*:363
Kroll process
  use of titanium dichloride, *24*:256
Krypton
  diffusion in vitreous silica, *21*:1048
Kyanite [*1302-76-7*], *21*:54
  refractories from, *21*:49
KYON, *24*:432

L

Laboratory information management
    systems
  use in technical service, *23*:779
Lac, *21*:299
Lacquers
  sampling standards for, *21*:627
  vinylidene chloride copolymer, *24*:915

Lac sulfur, *23*:257
L-Lactide [*95-96-5*], *23*:546
Lactisole [*13794-15-5*], *23*:576
Lactitol [*585-86-4*], *23*:95
  as sweeteners, *23*:556
Lactobacillus brevis
  inhibited by sorbates, *22*:580
Lactomer, *23*:547
  as sutures, *23*:542
Lactose [*63-42-3*], *23*:90
  compared to sucrose, *23*:4
Ladenburg's formula, *22*:633
Ladle metallurgy
  role in steelmaking, *22*:780
Ladles
  refractories for, *21*:83
LAER. See *Lowest achievable emission
    rate.*
Laksholic acid, *21*:300
Lamb wave, *21*:823
Laminar flow, *21*:348
Laminated safety glass, *24*:932
Laminates
  for cooling tower packing, *22*:213
  silylating agents for, *22*:152
Lamps
  silica in, *21*:1000
  vitreous silica in, *21*:1066
Landfill gas
  as fuel, *21*:231
  fuel for steam prdn, *22*:746
Landfill leachates
  reverse osmosis of, *21*:323
Land reclamation, *23*:740
Lane and Eynon constant volume
    procedure
  test for reducing sugars, *23*:15
Langmuir adsorption, *21*:319
Langmuir-Blodgett films, *23*:1077
Lang's lay cord, *24*:170
Lanolin [*8006-54-0*]
  membrane fouling, *24*:612
  as release agent, *21*:210
Lanolin alcohols, *23*:518
Lanosterol [*79-63-0*], *22*:870; *23*:874
Lantern mantles
  thorium for, *24*:71
Lanthanum arachidate
  thin films of, *23*:1081
Larch wood, *23*:98
Lard
  in soap manufacture, *22*:302

Lasalocid [25999-31-9], 24:831
Laser diodes
    compound semiconductors, 21:789
Laser dye
    stilbene dyes in, 22:929
Laser-induced breakdown spectroscopy
    (libs), 22:647
Laser-induced fluorescence, 22:653
Laser mirror
    tool materials for, 24:436
Laser printers
    use of selenium, 21:709
Lasers
    measurement of particles in, 22:275
    in optical spectroscopy, 22:635, 638,
        643
    use of selenium, 21:715
Laser sensors
    use of frequency modulated absorption,
        22:658
Laser surgery
    vitreous silica fibers in, 21:1069
Laser welding
    vitreous silica fibers in, 21:1069
Latex, 21:483, 532, 563
    use of ultrafiltration, 24:622
    from VDC copolymers, 24:916
Latex acrylics
    as sealants, 21:660
Latex allergies, 21:585
Latex resins
    thioglycolic acids in, 24:16
Lathers
    amine soaps in, 22:324
Lattice fluid hydrogen bonding model
    for supercritical fluids, 23:463
Laumontite
    silylating agents for, 22:147
Lauric acid [143-07-7]
    as rubber chemical, 21:473
    separation coefficients for, 21:926
Lauryl alcohol
    sulfation of, 23:173
Lauryl sulfate [151-21-3]
    wetting agents, 21:479
Laurylthiopropionic acid [1462-52-8], 24:2
Lava deposits, 21:54
Law of mass action
    in thermodynamics, 23:1024
Laws of thermodynamics, 23:985
Laxation thresholds
    for sugar alcohols, 23:107

LC-Fining process, 23:739
Leached glass fibers, 21:123
Leaching
    of ores using sulfur, 23:260
    of pesticides, 22:444
Lead [7439-92-1]
    analysis by x-ray absorption, 22:655
    in brazing filler metals, 22:492
    plating with methanesulfonic acid,
        23:326
    regulatory level, 21:160
    in scrap for steel, 22:779
    sulfuric acid as by-product, 23:381
    tellurium as by-product, 23:786
    tellurium in ores, 23:782
    thiosulfate complexes, 24:55
    vapor pressure of, 23:1046
Lead acetate
    for hydrogen sulfide detection, 23:282
Lead–acid batteries
    precipitated silica in, 21:1025
Lead alloys
    as solders, 22:485
    strontium in, 22:950
Lead antimony alloys
    use of selenium in, 21:711
Lead diamyldithiocarbamate [36501-84-5],
    21:466
Lead dimethyldithiocarbamate [19010-
    66-3], 21:466
Lead salicylate [15748-73-9], 21:612
Lead–selenium alloys, 21:711
Lead–silver
    solders, 22:487
Lead–sodium alloys, 22:347
Lead stearate [7428-48-0], 21:207
Lead–tellurium alloys, 23:803
    thermoelectric material, 23:1034
Lead thiosulfate [26265-65-6], 24:53
Lead–tin–silver alloys
    as solders, 22:487
Lead titanate [12060-00-3], 24:254
Lead titanate zirconate [12626-81-2],
    24:254
Lead–zinc
    silver as by-product, 22:171
Leaking underground storage tanks,
    21:166
Leather
    sampling standards for, 21:627
    sodium sulfides for prdn, 22:414
    sulfonic acid-derived dyes in, 23:206,
        207

sulfur dyes for, *23*:361
  thioglycolic acids for, *24*:15
  triarylmethane dyes for, *24*:565
  water resistance for, *22*:47
Leather dyes
  stilbene dyes, *22*:928
Leather tanning
  sodium sulfite used in, *23*:314
LED. See *Light-emitting diode.*
Lefebvre's equation, *22*:683
Legg cutter, *23*:758
Legionnaires' disease, *23*:979
Lemon
  for tea, *23*:762
Length
  measurement of, *24*:628
Lenses
  coatings for, *21*:1023
  refractory coatings for, *21*:113
  selenium in infrared, *21*:712
  of vitreous silica, *21*:1065
LEPC. See *Local Emergency Planning
    Committee.*
Lepidolite [*1317-64-2*]
  rubidium in, *21*:591
Les, *23*:1036
Leucine
  in silk, *22*:156
  in soybeans and other oilseeds, *22*:596
Leucine α-ketoglutarate transaminase
  in teas, *23*:751
Leuco bases
  of triarylmethane dyes, *24*:558
Leuco Sulfur Black 1 [*66241-11-0*], *23*:343
Leuco Sulfur Brown 96, *23*:353
Leuco Sulfur Red 14, *23*:343
Leucoxene [*1358-95-8*], *24*:241
Leucrose [*7158-70-5*], *23*:8, 25, 75
Levamisole [*14769-73-4*], *24*:830
Levans, *23*:8
  in beet sugar, *23*:55
Lever 2000, *22*:323
Lever rule, *21*:930
Levopimaric acid [*79-54-9*], *21*:294
Levulinic acid [*123-76-2*]
  dextrose degradation product, *23*:585
Levulose, *23*:87, 594
Lewis-Randall rule, *23*:1014
Lewis relation, *22*:206
Lewmet 55, *23*:393
Libs. See *Laser-induced breakdown
    spectroscopy.*

LICA 38, *24*:330
Lichens
  mannitol in, *23*:97
Licorice
  sweeteners from, *23*:568
Lidocaine [*137-58-6*], *24*:835
Lidofilcon-B, *24*:1079
Liebigite [*14831-68-6*], *24*:643
Liesegang rings, *22*:498
Lifetimes
  fluorescence, *22*:653
Light-emitting diode (LED)
  compound semiconductor, *21*:788
Light-polarizing films
  stilbene dyes in, *22*:929
Light scattering
  for sol–gel technology studies, *22*:507
Light sources
  for optical spectroscopy, *22*:636
Lightwave guides
  titanium coatings, *24*:323
Lignins
  extraction of, *23*:138
  separation by reverse osmosis, *21*:327
  sulfonated, *23*:205
  vanillin from, *24*:814
  water additive of steam prdn, *22*:740
Lignone, *21*:694
Lignosol SFX, *23*:619
Lignosulfates, *23*:205
Lignosulfonates, *23*:167, 495
  for tall oil recovery, *23*:619
Lilac
  extraction using SCFs, *23*:466
  mannitol in, *23*:97
Lime
  for reverse osmosis pretreatments,
    21:315
  sampling standards for, *21*:627
Lime–alumina–iron–silica, *21*:47
Lime–limestone processes
  FGD systems, *23*:447
Lime softening
  water treatment for steam prdn, *22*:745
Limestone
  centrifugal separation of, *21*:861
  role in steelmaking, *22*:767
  role in sulfur prdn, *23*:240
  sampling standards for, *21*:627
Liming
  of sugar beet juice, *23*:51

Limiting viscosity number, *21*:357
Limonene [*5989-27-5*], *23*:834
Ψ-Limonene [*499-97-8*], *23*:847
(+)-Limonene [*5989-27-5*], *23*:844
Limonin [*1180-71-8*], *23*:571
Limonite
    magnetic intensity, *21*:876
Linalool [*78-70-6*], *23*:835
Linalyl acetate [*115-95-7*], *23*:856
Lincomycin [*154-21-2*], *24*:829
Lindane [*58-89-9*], *22*:421, 934
    regulatory level, *21*:160
Linear alkylbenzene
    sulfonation of, *23*:158
Linear alkylbenzene sulfonate, *23*:158
Linear ethoxylates
    sulfation of, *23*:170
Linear sensor arrays
    amorphous semiconductors in, *21*:750
Linear superelasticity
    of shape-memory alloys, *21*:968
Linear units, *24*:636
Linewidths
    role in optical spectroscopy, *22*:632
Linings
    refractories for, *21*:84
Linoleic acid [*60-33-3*], *23*:617; *24*:504
    in soybeans and other oilseeds, *22*:608
Linolenic acid [*463-40-1*], *23*:617
    in soybeans and other oilseeds, *22*:608
Linseed oil
    boiling point of, *23*:627
    specific gravity, *23*:625
Lipids
    in soybeans and other oilseeds, *22*:596
Lipooligosaccharide
    of bacteria, *24*:736
Lipoxygenase
    in soybeans, *22*:596
Liquefaction of gases, *21*:128
Liquefied natural gas
    tanks for, *23*:644
Liquid ammonia, *21*:138
Liquid atomizers, *22*:670
Liquid chromatography/mass
        spectrometry
    for trace and residue analysis, *24*:504
Liquid crystal displays
    amorphous semiconductors in, *21*:750
    glass, *22*:79
Liquid crystalline polyurethanes, *24*:701
Liquid crystals, *22*:149
    p-hydroxybenzoic acid in, *21*:620

Liquid-injection-molded rubber, *22*:112
Liquid invert, *23*:41
Liquid-nitrogen traps, *24*:775
Liquid-ring pumps, *24*:777
Liquids
    sampling of, *21*:638
    use in thermometers, *23*:829
Liquid soaps, *22*:322
Liquid sucrose, *23*:41
Litharge [*1317-36-8*]
    as rubber chemical, *21*:473
Lithium [*7439-93-2*], *24*:854
    batteries, *23*:140
    in brazing filler metals, *22*:494
    complexes with sugar alcohols, *23*:103
    diffusion coefficient in vitreous silica,
        21:1047
    vitreous silica impurity, *21*:1036
Lithium acrylate [*13270-28-5*]
    in grouts, *22*:455
Lithium aluminum hydride, *22*:58
    for chlorosilanes, *22*:44
Lithium batteries
    silver cmpds in, *22*:190
    use of selenium, *21*:715
Lithium bromide
    for silk processing, *22*:159
Lithium chloride
    solubility in steam, *22*:728
Lithium dititanate [*12600-48-5*], *24*:251
Lithium fluoride [*7789-24-4*]
    transmission limit, *22*:643
    use in x-ray spectroscopy, *22*:654
Lithium hydride, *22*:59
    for chlorosilanes, *22*:44
Lithium hydroxide
    water additive of steam prdn, *22*:740
Lithium metatitanate [*12031-82-2*], *24*:251
Lithium orthotitanate [*12768-28-4*],
        24:251
Lithium reservoir
    titanium sulfide as, *24*:266
Lithium salicylate [*552-38-5*], *21*:612
Lithium silicates [*12627-14-4*], *22*:1
Lithium–sodium alloys, *22*:347
Lithium tantalate [*12031-66-2*], *23*:676
Lithium tetrahydridothallate(III)
        [*82374-47-8*], *23*:956
Lithium thiocyanate
    for silk processing, *22*:159
Lithography
    compound semiconductors, *21*:809
    thin films for, *23*:1084

Living polymerization
    solution SBR by, 22:1005
    of VE monomers, 24:1058
Living styrene polymerization, 22:1045
Local Emergency Planning Committee
        (LEPC), 21:166
Lo-Cat process, 23:444
Logarithmic viscosity number, 21:356
Logwood extract, 23:550
Londax
    thiols in, 24:30
Long-acting thyroid stimulator [9034-48-
        4], 24:89
Lonzaine
    surfactant, 23:531
Lorandite [15501-93-6], 23:952
Lorenzenite, 24:264
Lornoxicam [70374-39-9], 24:47
Lovibond Comparator
    rubber color, 21:568
Loweite, 22:404
Lowest achievable emission rate (LAER),
        21:166
Low pressure CVD, 23:1060
Lube additives, 23:163
Lubricants
    additives, 23:430
    antifriction, 23:212
    organotin catalysts for, 24:145
    refractory coatings as, 21:113
    in refrigeration systems, 21:147
    rheological measurements, 21:427
    selenium in, 21:713
    sodium tetrasulfide in mfg of, 22:418
    strontium compounds as, 22:954
    sulfonates in, 23:164
    tanks for, 23:644
    thioglycolic acids in, 24:16
    titanate catalysis, 24:325
Lubricating oils
    purification by centrifuge, 21:856
    recycling of, 21:2
    sulfur additive, 23:291
Ludox spheres, 22:499
Ludox TM50, 21:1008
Luff Schoorl
    reducing sugar test, 23:16
Lumisterol [474-69-1], 22:857
Lunar caustic, 22:183
Lungworms
    veterinary drugs against, 24:830
Lupus erythematosis, 24:837

Luteinizing hormone [9002-67-9], 24:833
Luttinger parameters, 21:768
Luviflex, 24:1094
Luviquat, 24:1094
Luviskol, 24:1094
Lux, 22:323
Lycasin, 23:25
Lycopene [502-65-8], 23:874
Lycra, 24:718
Lyell syndrome, 24:515
Lyral [31906-04-4], 23:849
Lysine
    in silk, 22:156
    in soybeans and other oilseeds, 22:596
Lysozyme crystals
    grown in space, 22:622
Lyxitol [488-82-4], 23:96

M

M19001, 22:491
Mace, 21:259
Mackenzie-Shuttleworth model, 22:519
MacMichael viscometer, 21:397
MACT. See Maximum achievable control
        technology.
Magadiite [12285-88-0], 22:5
Magdol
    refractories, 21:83
Maghemite
    magnetic intensity, 21:876
Magic Acid, 23:209
Magnelli phases, 24:234
Magnesias
    refractories from, 21:50
Magnesioferrite, 21:56
Magnesite [13717-00-5], 21:55
    refractories from, 21:50
Magnesium
    in brazing filler metals, 22:487
    complexes with sugar alcohols, 23:103
    ferrosilicon in prdn of, 21:1114
    sulfur removal, 22:781
    in tea, 23:751
    vitreous silica impurity, 21:1036
Magnesium acrylate [5698-98-6]
    in grouts, 22:455
Magnesium alloys
    brazing filler metals, 22:491
    thorium in, 24:71
    use of selenium in, 21:711

Magnesium carbonate [546-93-0], 22:767
  salt additive, 22:368
Magnesium chloride, 22:365
  catalyst in textile finishing, 23:898
  in steam cooling water, 22:744
  use in refractory brick, 21:69
Magnesium dititanate [12032-35-8], 24:252
Magnesium diuranate [13568-61-1], 24:650
Magnesium ferrosilicon, 21:1117
Magnesium fluoride, 23:1046
Magnesium hydride [7693-27-8], 22:59
Magnesium hydroxide
  thickening of, 21:680
Magnesium metatitanate [1312-99-8], 24:252
Magnesium orthotitanate [12032-52-9], 24:252
Magnesium oxide [1309-48-4]
  as refractory material, 21:57
  as rubber chemical, 21:473
  sampling standards for, 21:627
  solubility in steam, 22:728
Magnesium salicylate tetrahydrate [18917-89-0], 21:612
Magnesium silicide, 22:43
Magnesium–sodium alloys, 22:347
Magnesium stannate [12032-29-0], 24:128
Magnesium stearate [557-04-0], 22:324
  as release agent, 21:210
Magnesium sulfamate, 23:131
Magnesium sulfate, 22:365
  use in refractory brick, 21:69
Magnesium thiosulfate [10124-53-5], 24:64
Magnet alloys
  tellurium in, 23:803
Magnetic saturation
  role in tool coatings, 24:423
Magnetic semiconductors, 21:763
Magnetic tape
  refractory coatings for, 21:113
Magnetic wire insulation
  PVF resins, 24:938
Magnetite, 21:56
  magnetic intensity, 21:876
Magnetos
  engine, 21:342
Magnetrons, 22:636
Magnetron sputtering
  for thin films, 23:1053
Magnets, 24:327
  permanent ceramic, 22:951

selenium in, 21:711
for separation, 21:884
for thin-film formation, 23:1053
vanadium in, 24:795
Mahogany acids, 23:497
Mahogany sulfonate, 23:163
Maillard reaction, 23:310, 585
  aspartame in, 23:560
Maize
  dextrose from, 23:585
Makatite [27788-50-7], 22:5
Makon
  ethoxylated alkylphenol surfactant, 23:511
Malachite green [569-64-2], 24:553
Malaria, 21:238
  developing vaccines for, 24:735
Maleic anhydride [108-31-6]
  sampling standards for, 21:627
  styrene copolymers, 22:1024
  succinic anhydride from, 22:1081
  VP copolymerization, 24:1090
Malonic acid [141-82-2]
  sorbic acid from, 22:575
  from succinic acid, 22:1077
Maltitol [585-88-6], 23:90, 95
  as sweeteners, 23:556
Maltitol solutions, 23:100
Maltitol syrups, 23:100
Maltodextrin [9050-36-6], 23:25, 582, 598
  aspartame in, 23:559
  with sucralose, 23:569
Maltose [69-79-4], 23:89
  compared to sucrose, 23:4
  in corn syrup, 23:598
Maltose syrup [69-79-4], 23:25, 599
Maltotriose
  in corn syrup, 23:598
Malt sugar, 23:89
Malt syrups, 23:90
Malvalic acid, 22:609
  in cottonseed, 22:598
Management
  of research and technology, 21:264
Manganese
  in brazing filler metals, 22:492
  emissivities of, 23:827
  in ferrous shape-memory alloys, 21:965
  in pig iron, 22:767
  in steel, 22:775
  in tool steels, 24:398
  vitreous silica impurity, 21:1036

Manganese–bronze
  for acid hydrolysis containers, 23:599
Manganese nodules
  tellurium in, 23:782
Manganese–silicon alloys, 21:1118
Manganese(II) sulfide [18820-29-6]
  in steel, 22:815
Mange
  selenium cmpds for, 21:713
Manila copal [9000-14-0], 21:297
Manjak, 21:299
Manna ash
  mannitol in, 23:97
Mannheim process, 22:406; 23:259
Mannich reaction, 23:899
Mannitol [69-65-8]
  as sweeteners, 23:556
D-Mannitol [69-65-8], 23:95
D,L-Mannitol [133-43-7], 23:95
L-Mannitol [643-01-6], 23:95
Maple butter, 23:601
Maple cream, 23:601
Maple sugar, 23:601
Maple syrup, 23:582, 601
Maprosyl 30
  surfactant, 23:493
Maprotiline [10262-69-8], 22:942
Maraschino cherries
  sorbates in, 22:583
Marcasite
  selenium in, 21:689
Margarine
  soybeans and other oilseeds in, 22:610
  trace analysis of, 24:515
  use of sorbates, 22:582
Margules equation, 21:390
Marignac process, 23:661, 675
Marine algae
  mannitol in, 23:97
Marine equipment
  sulfamic acid cleaner for, 23:129
Marine piping systems
  shape-memory alloys for, 21:969
Marine steroids, 22:852
Mark-Houwink relation, 21:358
Markovnikov's rule, 23:160
Markownikoff addition
  for thiols, 24:22
Marplan, 22:940
Marshmallows
  corn syrups in, 23:600
Marsilid, 22:940

Martempering
  of steel, 22:804
Martensite, 22:791
  phase properties of, 22:798
Martin diameter, 22:266
Martite
  magnetic intensity, 21:876
Mason number, 21:366
Mass spectrometer
  rhenium filaments for, 21:337
Mass spectrometry
  atomic spectroscopy contrasted, 22:645
  use of silylating agents in, 22:143
Mass transfer
  with simultaneous heat transfer, 22:196
Mastic [61789-92-2], 21:298
Mastitis, 24:837
Material Safety Data Sheet, 21:166
Materials of construction
  in sulfuric acid manufacture, 23:393
Materials science
  analysis by Mössbauer spectroscopy,
    22:656
Mathcad
  for reverse osmosis models, 21:310
Mats
  in reinforced plastics, 21:195
Maxillofacial surgery
  shape-memory alloys for, 21:973
Maximum achievable control technology
    (MACT), 21:166; 22:532
Maxon, 23:543
Maxwell equations, 23:997
MCM-22
  zeolite catalyst, 22:966
McMurry reaction, 24:306
MDI. See 4,4'-Methylenebis(phenyl
    isocyanate).
Mean field theory of polymerization,
    22:507
Measles
  vaccines against, 24:727
Meat
  centrifugal separation of, 21:861
  dextrose in, 23:592
  lipid removal from, 23:467
  use of sorbates, 22:583
Meat curing
  sodium nitrite for, 22:394
Meat products
  extraction of lipids, 23:467
Mebanazine [65-64-5], 22:940

Mebendazole [31431-39-7], 24:831

Mechanical interlocking
  binder, 22:224

Medical Device Amendments of 1976,
  23:553

Medical Device Reporting, 23:541, 554

Medical devices
  regulatory agencies, 21:175

Medroxyprogesterone [520-85-4], 22:903

Meewein-Ponndorf reaction, 24:284

Mefloquine, 24:508

Megestrol acetate [3562-63-8], 24:833

Megestrol acetate [595-33-5] (in humans),
  22:903

Melamine–formaldehyde
  in textile finishing, 23:896

Melamines
  silane coupling agent for, 22:152
  in textile finishing, 23:898

Melanophlogite, 21:988

Melphalan [148-82-3], 24:836

Melt spinning, 23:885

Melt viscosity, 21:359

Membranes, 21:303
  gas permeation, 23:462
  precipitated silica in, 21:1025
  for roofing, 21:446
  sulfonic acid derivatives for, 23:169
  for ultrafiltration, 24:603

Membrane technology
  for sodium chloride prdn, 22:366

Meningitis
  vaccines against, 24:728

p-2,4(8)-Menthadiene [138-86-3], 23:834

p-3,8-Menthadiene [586-67-4], 23:834

cis-p-Menthane [6069-98-3], 23:834, 845

trans-p-Menthane [1678-82-6], 23:834, 845

cis-p-Menthane-3,8-diol [92471-23-3],
  23:867

trans-p-Menthane-3,8-diol [91739-72-9],
  23:867

cis/trans-p-8-Menthene [6252-33-1], 23:834

p-1-Menthene [499-94-5], 23:834

p-3-Menthene [500-00-5], 23:834

p-4(8)-Menthene [34105-55-0], 23:834

trans-2-p-Menthene [1124-26-1], 23:834

Menthol [1490-04-6], 23:835

(±)-Menthol [15356-70-4], 23:858

(−)-Menthol [2216-51-5], 23:858

trans-Menthone [89-8-5], 23:835

Menthyl salicylate [89-46-3], 21:614

Mepivacaine hydrochloride [1722-62-9],
  24:835

Mercaptans, 24:19
  removal of, 23:436
  sulfurization of, 23:429

Mercaptoacetic acid [68-11-1], 24:1, 21

S-Mercaptoacetylthioacetic acid [99-68-3],
  24:2

o-Mercaptobenzoic acid [147-93-3], 21:623

2-Mercaptobenzothiazole [149-30-4],
  21:465

Mercaptobenzothiazoles
  vulcanization accelerator, 21:461

2-Mercaptoethanol [60-24-2], 24:21

2-Mercaptopropionic acid [79-42-5], 24:6, 7

3-Mercaptopropionic acid [107-96-0], 24:2,
  6, 7, 21

3-Mercaptopropyltrimethoxysilane [4420-
  74-0]
  adhesion promoter, 21:656
  critical surface tension of, 22:148
  silane coupling agent, 22:150

2-Mercapto-1,3,4-thiadiazole-5-thioben-
  zoate [51988-14-8]
  cross-linking agent, 21:470

Mercuration
  of toluene, 24:357

Mercuric chloride [7487-94-7], 24:852

Mercuric iodide crystals
  by physical vapor transport, 22:625

Mercury [7439-97-6]
  boiling point of, 23:627
  impurity in selenium, 21:705
  regulatory level, 21:160
  for sodium prdn, 22:339
  thiosulfate complex, 24:55
  use in thermometers, 23:829
  vapor pressure of, 23:1046
  vitreous silica impurity, 21:1036
  vitreous silica windows, 21:1068

Mercury–cadmium–telleride
  photodiodes for optical spectroscopy,
  22:642

Mercury salicylate [5970-32-1], 21:612

Mercury–sodium alloys, 22:347

Mercury telluride [12068-90-5], 21:767

Mercury–thallium alloy [83542-96-5],
  23:954

Mercury vapor lamps
  vitreous silica sealings, 21:1041

Merkel's approximation, 22:213

Merphos
  thiols in, 24:30

Merrifield resin, 22:150

Mersilene, 23:543
Mesityl oxide [141-79-7]
    as solvent, 22:546
Mesogenic diols
    for liquid crystalline polyurethanes,
        24:701
Mestranol [72-33-3], 22:903
Metabolism
    of pesticides, 22:425
Metakahlerite [12255-22-0], 24:643
Metal coatings
    tellurium in, 23:803
Metal conditioners
    PVB in, 24:935
Metal insulator semiconductor field-effect
        transistor (MISFET), 21:778
Metallic coatings
    sampling standards for, 21:627
Metallic pinwheels
    as bird repellents, 21:254
Metallic soaps, 22:324
    as release agents, 21:207
Metallides, 21:88
Metalliding, 21:106; 23:1072
Metallocenes
    thallium cmpds in prprn of, 23:959
Metallurgy
    analysis by Mössbauer spectroscopy,
        22:656
    of steel, 22:780
    tellurium as machining additive,
        23:790
    of tin, 24:109
Metal organic chemical vapor deposition
        (MOCVD), 21:764, 769; 23:1060
Metal oxides
    as foulants for reverse osmosis, 21:313
    as sensors, 21:825
Metal oxide semiconductor field-effect
        transistor (MOSFET), 21:720, 778
Metal pickling
    sodium bisulfate for, 22:411
Metals
    analysis by atomic emission
        spectroscopy, 22:647
    in compound semiconductor processing,
        21:804
    separation by reverse osmosis, 21:322
    trace analysis of, 24:517
Metals cleaning
    use of sulfamic acid, 23:129
Metal semiconductor field-effect
        transistor, 21:779

Metals finishing
    sodium tetrasulfide in, 22:418
Metal shingles
    on roofs, 21:453
Metal stearates
    as release agents, 21:210
Metal surface treatments
    thioglycolic acid for, 24:15
Metal treatments
    use of titanium compounds, 24:267
Metam sodium, 22:422
Metasulfuron, 24:510
Metatitanic acid [12026-28-7], 24:234
Methacrylates
    for hydroseeding, 22:461
Methacrylic acid
    VP copolymerization, 24:1090
3-Methacryloxypropyltrimethoxysilane
        [2530-85-0]
    adhesion promoter, 21:656
    silane coupling agent, 22:150
Methamphetamine [537-46-2], 22:937
Methane [74-82-8], 24:855
Methanesulfonic acid [75-75-2], 23:194,
        325
Methanesulfonyl chloride [124-63-0],
        23:323
Methanethiol [74-93-1], 24:21
Methanol [67-56-1], 23:440
    dielectric constants of, 22:759
    from solar energy, 22:480
    as solvent, 22:536
    supercritical fluid, 23:454
    supercritical fluid cosolvent, 23:457
Methimazole [60-56-0], 24:99
Methiocarb [2032-65-7]
    bird repellent, 21:256
Methionine, 22:608
    in silk, 22:156
    in soybeans and other oilseeds, 22:596
D,L-Methionine
    thiols for, 24:30
Methionine sulfoxide [454-41-1], 23:217
Methotrexate [59-05-2], 24:836
Methoxychlor [72-43-5]
    regulatory level, 21:160
4,6-O-(1-Methoxyethylidene) sucrose
        [116015-72-6], 23:69
Methoxyflurane [76-38-0], 24:835
3-Methoxy-4-hydroxy-benzaldehyde
        [121-33-5], 24:812
(17α)-3-Methoxy-19-norpregna-1,3,5(10)-
        trien-20-yne-17-ol, 22:903

2(4-Methoxyphenoxy)propionic acid, 23:576

p-Methoxyphenyl vinyl ether [4024-19-5], 24:1056

Methyl acetate [79-20-9]
    as solvent, 22:540

Methyl acetoacetate [105-45-3]
    as solvent, 22:542

Methyl acrylate
    copolymerization with VDC, 24:888

Methyl alcohol [67-56-1]
    as HAP compound, 22:532

Methyl 3-amino-4-methyl thiophene-2-carboxylate [85006-31-1], 24:44

N-Methylaminopropyltrimethoxysilane [3069-25-8]
    for liquid crystals, 22:149

Methyl 3-aminothiophene-2-carboxylate [22288-78-4], 24:3, 44

4-Methyl-2-amyl alcohol [108-11-2]
    as solvent, 22:536

Methyl amyl ketone [110-43-0]
    as solvent, 22:546

Methyl anthranilate [134-20-3]
    as bird repellent, 21:255
    saccharin starting material, 23:566

N-Methyl-D-aspartic acid [6384-92-5], 22:935

Methyl bromide [74-83-9], 22:422

2-Methyl-1-butanethiol [1878-18-8], 24:20

3-Methyl-1-butanol [123-51-3]
    as solvent, 22:536

4-(3-Methyl-2-butenyl)phenylmagnesium chloride [106364-41-4], 24:854

4-(3-Methyl-2-butenyl)styrene [85964-33-6], 24:854

Methyl t-butyl ether [1634-04-4], 22:480
    as HAP compound, 22:532

Methyl n-butyl ketone [591-78-6]
    as solvent, 22:546

3-Methyl-1-carbethoxy-2-thioimidazoline [22232-54-8], 24:100

Methylchavicol [104-67-0], 23:836

Methyl chloride [74-87-3], 24:865

Methylchlorosilane reaction, 22:84

Methyl chlorosilanes
    role in sol–gel technology, 22:515

Methyl chlorosulfate [812-01-1], 23:410

Methyl chlorosulfite [13165-72-5], 23:411

1-((6-Methyl-3-cyclohexen-1-yl)carbonyl)-pyrrolidine [67013-95-0]
    as repellant, 21:249

1-((2-Methylcyclohexyl)-carbonyl)-pyrrolidine [52736-60-4]
    as repellant, 21:249

Methylcyclopentadienylmanganese tricarbonyl, 22:344

Methyldichlorosilane [75-54-7], 22:38, 47

Methyldiethanolamine [105-59-9]
    for hydrogen sulfide removal, 23:437

Methyldisilane [13498-43-6], 22:43

Methylenebisacrylamide [110-26-9]
    in grouting systems, 22:454

4,4'-Methylenebis(cyclohexylamine) carbamate [15484-34-1]
    cross-linking agent, 21:470

4,4'-Methylenebis(phenyl isocyanate) (MDI) [101-68-8], 24:697, 706
    in grouts, 22:456

Methylene Blue [61-73-4], 23:346

Methylene chloride [75-09-2], 24:865
    for decaffeination of tea, 23:762
    as solvent, 22:538

Methylene chloride–2-propanol–water phase equilibrium, 21:932

4,4'-Methylene dianiline [101-77-9]
    cross-linking agent, 21:470

Methylenediphenyldiisocyanate [101-68-8]
    in urethane sealant, 21:657

Methylene-5,5'-disalicylic acid [122-25-8], 21:606, 616

Methylene dithiocyanate [6317-18-6], 23:323

Methylene sulfate [20757-83-9], 23:410

Methylene Violet [2516-05-4], 23:346

Methyl ethyl ketone [78-93-3], 21:930
    as HAP compound, 22:532
    regulatory level, 21:160
    as solvent, 22:546

Methylethylketoxime
    oxygen scavenger for steam, 22:742

Methyl fluorosulfate [421-20-5], 23:410

α-Methylheptenone [10408-15-8], 23:837

β-Methylheptenone [110-93-0], 23:836

6-Methylhept-5-en-2-one [110-93-0], 23:837

6-Methylhept-6-en-2-one [10408-15-8], 23:837

Methyl hydrogen sulfate [75-93-4], 23:410

α-n-Methylionone [1322-70-9], 23:865

γ-Methylionone [7779-30-8], 23:865

Methyl isoamyl ketone [110-12-3]
    as solvent, 22:546

Methyl isobutyl ketone [108-10-1]
  as HAP compound, 22:532
  as solvent, 22:546
  solvent extraction of tantalum, 23:661
  for thorium recovery, 24:70
Methyl isopropyl ketone, 21:930
Methyllithium
  reactions with rhenium, 21:345
Methylmalonic acid [376-05-2]
  in serum, 24:507
1-Methyl-2-mercaptoimidazole [60-56-0],
  24:100
Methyl 3-mercaptopropionate [2935-90-2],
  24:7
Methylmercury
  in fish, 24:514
Methyl methacrylate
  copolymerization with VDC, 24:888
  use of sulfuric acid, 23:397
  VP copolymerization, 24:1090
N-Methylolacrylamide [924-42-5]
  in grouting systems, 22:454
Methylol phenol–formaldehyde resin
  [26678-93-3]
  cross-linking agent, 21:470
6-Methyl-1,2,3-oxathiazin-4(3H)-one
  2,2-dioxide, 23:563
2-Methyl-n-pentane-2,4-diol, 24:289
4-Methyl-2-pentanol, acetate [108-84-9]
  as solvent, 22:542
Methylphenidate [113-45-1], 22:937
p-Methylphenyl vinyl ether [1005-62-5],
  24:1056
2-Methyl-1-piperidyl-3-cyclohexene-
  carboxamide [69462-43-7], 21:246
Methylpolysiloxane, 22:147
6α-Methylprednisolone [83-43-2], 22:905
2-Methyl-1-propanethiol [513-44-0], 24:21
2-Methyl-2-propanethiol [75-66-1], 24:21
2-Methyl-1-propanol [78-83-1]
  as solvent, 22:536
2-Methyl-2-propanol [75-65-0]
  as solvent, 22:536
Methyl propionate [554-12-1]
  as solvent, 22:542
Methyl n-propyl ketone [107-87-9]
  as solvent, 22:546
Methylpseudoionone [26651-96-7], 23:865
N-Methyl-2-pyrrolidinone [872-50-4]
  as solvent, 22:546
Methyl salicylate [119-36-8], 21:601, 614
Methylsilane [992-94-9], 22:38

p-Methylstyrene [627-97-9], 22:987
α-Methylstyrene [98-83-9], 22:990
3-(Methylsulfinyl)alanine [4740-94-7],
  23:217
Methylsulfinylbenzene [1193-82-4], 23:218
Methylsulfinyl carbanion [15590-23-5],
  23:222
1-Methylsulfinyl-2-ethyl-3-phenyl propane
  [14198-15-3], 23:221
Methyltellurium tribromide [20350-53-2],
  23:800
Methyl-2-thiazolidinethione, 21:468
Methyl thioglycolate [2365-48-2], 24:3, 7
2-Methylthiophene [554-14-3], 24:36, 37
3-Methylthiophene [616-44-4], 24:36, 37
3-Methyl-2-thiophenecarboxaldehyde
  [5834-16-2], 24:48
Methylthiophenes, 24:34
6-Methyl-2-thiouracil [56-04-2], 24:100
Methyl p-toluenesulfonate [23373-38-8],
  23:412
Methyltriacetoxysilane [4253-34-3], 22:72
  as silicone cross-linker, 21:655
Methyl23-trichlorosilyltricosanoate
  monolayers of, 23:1091
Methyltriethoxysilane [2031-67-6], 22:72
Methyl triflate [333-27-7], 23:412
Methyltrifluoropropyl silicone, 22:47
Methyltrimethoxysilane [1185-55-3],
  22:72
  critical surface tension of, 22:148
  as silicone cross-linker, 21:655
N-Methyl-N-trimethylsilyltrifluoro-
  acetamide [24589-78-4]
  silylating agent, 22:144
Methyltris(butanoneoxime)silane [22984-
  54-9]
  as silicone cross-linker, 21:655
N-Methyl-N-vinylacetamide [3195-78-6],
  24:1071
Methyl vinyl ether [107-25-5], 24:1054
Methyl violet [8004-87-3], 24:562
Methylxanthines
  as stimulants, 22:936
  in tea, 23:747
Metolachlor [51218-45-2], 22:422, 427
Metravib Micromecanalyser, 21:420
Metronidazole [443-48-1], 24:830
Mevalonic acid [150-97-0], 22:870
MGK R-874
  thiols in, 24:30
MGK Repellent 11 [126-15-8]
  as mosquito repellent, 21:244

MG rubber, 21:575
Mibolerone [3704-09-4], 24:833
Mica [12001-26-2], 21:980
    as release agent, 21:210
    sampling standards for, 21:627
    substrate for self-assembled monolayer,
        23:1089
    use in optical spectroscopy, 22:636
Michael additions, 24:24
Michler's hydrol [119-58-4], 24:555, 562
Michler's ketone [90-94-8], 24:560
Microbalances
    synthetic quartz crystals for, 21:1082
Microbial metabolism
    of pesticides, 22:425
Microbiological stains
    triarylmethane dyes for, 24:565
Micrococcus sp.
    inhibited by sorbates, 22:580
Microcrystalline wax [64742-42-3]
    as release agent, 21:210, 212
Microdensitometers
    use in atomic emission spectroscopy,
        22:646
Microemulsions
    sulfolane in, 23:140
    water-in-oil, 23:462
Microextraction
    in trace analysis, 24:493
Microfiltration, 24:603
    pretreatment for membrane feed,
        21:316
Microgravity, 22:620
Microlite [12173-96-5]
    tantalum in, 23:660
Micromachining
    for sensors, 21:820
    of silicon, 21:1089
Micromechanics
    silicon-based semiconductors, 21:744
Microprocessor chips, 21:720
Micro-Quartz, 21:123
Microscopy, 21:111
    use in particle size measurement,
        22:266
Microsensors
    chemical, 21:824
Microspheres
    from SCFs, 23:472
Microtome knives, 24:436
Microwave spectroscopy, 22:636

Microwave technology
    for diamond films, 24:439
    residue analysis of, 24:495
Middle soap, 22:303
Midge
    silk source, 22:156
Mie scattering, 22:628
Milbemycin, 24:830
Mild steel
    for tanks and pressure vessels, 23:643
Milk
    centrifugal separation of, 21:859
    extraction of lipids, 23:467
    use of ultrafiltration, 24:622
Milk glass, 24:323
Milk sugar, 23:90
Milled fibers
    in reinforced plastics, 21:195
Milled soap, 22:314
Millet
    protection from birds, 21:254
Millimeter wave-integrated circuits,
        21:778
Mineralocorticoids, 22:852
Mineral oil
    specific gravity, 23:625
Mineral processing
    sedimentation in, 21:682
Mineral recovery and processing
    silicates for, 22:22
Minerals, 23:956
    particle size measurement for, 22:270
    trace analysis of, 24:518
Mineral seal oil
    as solvent, 22:538
Mineral spirits
    as solvent, 22:538
Mining
    reverse osmosis for wastewater, 21:322
Mirabilite, 22:403
Miracle Fruit, 23:575
Miraculin [143403-94-5 or 125267-18-7],
        23:575
Mirror blanks
    of vitreous silica, 21:1068
Mirrors
    chemical silvering of, 23:1074
    silver cmpds for coatings, 22:176, 192
MISFET. See Metal insulator
        semiconductor field-effect transistor.
Miso, 22:615
Mistletoe
    mannitol in, 23:97

Miticides
  selenium cmpds as, *21*:713
Mixers, *22*:234
Mixing
  property changes in thermodynamics, *23*:1016
Mobile liquids
  binder, *22*:224
  particle-bonding mechanism, *22*:225
MOCVD. See *Metal organic chemical vapor deposition.*
Modified natural rubber, *21*:575
Modified starches, *22*:713
Modulation doped FET, *21*:780
Mogroside V [*88901-36-4*]
  potential sweetener, *23*:573
Mohs' hardness
  of celestite, *22*:949
  of α-quartz, *21*:1076
  sodium chloride, *22*:359
  of talc, *23*:611
  of tellurium, *23*:784
  titanium compounds, *24*:229
  titanium dioxide, *24*:236
Moisture-curing silicones, *21*:654
Moisture-releasing silicones, *21*:654
Molasses, *23*:25, 42, 602
  desugarization of, *23*:58
  sugar alcohols, *23*:96
  syrup, *23*:582
  tanks for, *23*:647
  use of sulfur dioxide in mfg, *23*:310
Molasses Act, *23*:602
Molded rubber products
  from ground rubber, *21*:35
Moldings
  PVC use, *24*:1040
Mold-release agents, *21*:207
  in styrene plastics, *22*:1058
Molds
  inhibited by sorbates, *22*:579
Molecular absorption spectroscopy, *22*:642
Molecular beam epitaxy, *23*:1049
  for pure silicon, *21*:1094
Molecular beams
  analysis by multiphoton ionization, *22*:657
  use in optical spectroscopy, *22*:657
Molecular modeling, *23*:462
Molecular sieves
  for hydrogen sulfide removal, *23*:435
  inorganic silicates, *22*:25

Molecular transport, *24*:764
Molinate
  thiols in, *24*:29
Mollier chart, *22*:722
Molten salt
  use in solar energy, *22*:471
Molten salt electrolysis
  of refractory coatings, *21*:92
Molybdenite
  rhenium as by-product of, *21*:336
Molybdenum, *21*:58
  emissivities of, *23*:827
  refractory coating for, *21*:105
  role in steelmaking, *22*:779
  thorium oxides, *24*:76
  in tool steels, *24*:398
  vapor pressure of, *23*:1046
Molybdenum carbide
  in refractory coatings, *21*:91
Molybdenum disilicide [*1317-33-5*], *21*:58
Molybdenum foil
  sealing to vitreous silica, *21*:1041
Molybdenum–rhenium alloys, *21*:339
Molybdenum silicide
  as refractory coating, *21*:105
Moments of inertia
  from optical spectroscopy, *22*:636
Monamid LIPA
  surfactant, *23*:520
Monamid LMA
  surfactant, *23*:520
Monamid S
  surfactant, *23*:520
Monamine
  diethanolamine surfactant, *23*:520
Monateric
  surfactant, *23*:531
Monatin [*146142-94-1*]
  potential sweetener, *23*:573
Monawet
  dialkyl sulfosuccinate surfactant, *23*:498
Monazite [*1306-41-8*]
  magnetic intensity, *21*:876
  thorium in, *24*:70
Monazoline
  surfactant, *23*:527
Monel, *21*:107
  for centrifuges, *21*:845
Monellin [*9062-83-3*]
  potential sweetener, *23*:573
Monensin [*17090-79-8*], *24*:829, 831

Monetasone [83919-23-7], 22:906
Monitors
    stack-gas, 22:638
Monoalkanolamides, 23:519
Monoamine oxidase
    inhibitions of, 22:939
Monoaminotriphenylmethanes
    as dyes, 24:551
Monobutyltin oxide [51590-67-1], 24:147
Monobutyltin sulfide [15666-29-2], 24:152
Monochloracetic acid [79-11-8], 24:3
Monochromators
    use in optical spectroscopy, 22:643
Monocryl, 23:543
Monoethanolamine [141-43-5]
    for hydrogen sulfide removal, 23:437
Monoethanolamine thioglycolate [126-
    97-6], 24:8
10-Monohydroperfluoroundecanoic acid
    monolayer [1765-48-6]
    release substrate, 21:212
Monohydroxybenzoic acids, 21:601
Monolayers, 23:1077
    polymerization of, 23:1094
Monomethyl sulfate [512-42-5], 23:414
Monomethyltin tris(isooctylmercapto-
    acetate) [56225-49-1], 24:148
Monoperoxysuccinic acid [3504-13-0],
    22:1077
Monophenol monooxygenase [tyrosinase],
    23:751
Monophenyl sulfate [4074-56-0], 23:415
Monosilane [7803-62-5], 21:1087; 24:854
Monosilicic acid [10193-36-9], 21:1010,
    1042
Monosilicic acid, dimerized [1343-98-2],
    22:505
Monosof, 23:543
Monoterpene, 23:833
Montmorillonite [1318-93-0]
    in sol–gel polymer hybrids, 22:526
Mooney viscometer, 21:393, 397
Morantel [20574-50-9], 24:48
Morpholine [110-91-8], 23:287
    solubility in steam, 22:730
    as solvent, 22:540
    water additive of steam prdn, 22:740
Morpholinoglucopyranosides, 23:72
4-Morpholinyl-2-benzothiazoledisulfide
    [95-32-9], 21:465
Mortars
    for refractories, 21:48
    silicates in prdn of, 22:23

MOS DRAM technology, 21:720
MOSFET. See Metal oxide semiconductor
    field-effect transistor.
Mosquito repellents, 21:240
Mössbauer spectroscopy, 22:656
Moss silver, 22:184
Mothproofing agents
    triarylmethane derivatives, 24:565
Motor carriers, 24:527
Motors
    steel for, 22:826
Mountain ash, 23:96
MQ resins, 22:115
MSW. See Municipal solid waste.
Mucor silvaticus
    inhibited by sorbates, 22:579
Mucor spp.
    inhibited by sorbates, 22:579
Müller's solution
    reducing sugar test, 23:16
Mullite [55964-99-3]
    fibers from sol–gel technology, 22:524
    refractories from, 21:50
    as refractory material, 21:57
Multilayer barrier film, 24:914
Multiphoton absorption spectroscopy,
    22:657
Multiphoton ionization
    of liquids and interfaces, 22:657
Mumps
    vaccines against, 24:727
Municipal solid waste (MSW)
    as renewable energy source, 21:227
Municipal water treatment
    use of reverse osmosis, 21:324
Muriatic acid
    specific gravity, 23:625
Muscovite
    magnetic intensity, 21:876
Mushrooms
    sugar alcohols in, 23:96
Muskeg, 23:730
Muskol, 21:241
Mussels
    trace analysis of, 24:515
Mustard oils
    sulfoxides in, 23:217
Mycobacteria
    destruction of, 22:847
Mycobacterium avium
    destruction of, 22:847
Mycobacterium bovis
    destruction of, 22:847

Mycobacterium tuberculosis
  destruction of, *22*:847
Mycoplasma sp.
  veterinary drugs against, *24*:829
Mycoses
  veterinary drugs against, *24*:829
Mylar, *24*:324
Mylar ribbons
  as bird repellents, *21*:254
Myrcene [*123-35-3*], *23*:834
Myrcenol [*543-39-5*], *23*:848
Myricetin [*529-44-2*]
  in tea, *23*:749
Myristic acid [*544-63-8*]
  separation coefficients for, *21*:926
Myrj
  polyoxyethylene surfactant, *23*:514
Myrothecium roridum
  inhibited by sorbates, *22*:579
Myrothecium sp.
  inhibited by sorbates, *22*:579
Myrothecium verrucaria
  inhibited by sorbates, *22*:579
Myrrh [*9000-45-7*], *21*:299
Myxedema
  role of the thyroid, *24*:91

N

NAAQS. See *National Ambient Air
    Quality Standard.*
NaDBC. See *Sodium di-n-butyldi-
    thiocarbamate.*
Nafion-H [*66796-30-3*], *23*:210
Nagyagite [*12174-01-5*]
  tellurium in, *23*:783
Nalorphine [*62-67-9*], *24*:835
Naloxone [*465-65-6*], *24*:835
Nametre viscometer, *21*:403
Nanofiltration, *21*:323, 327
Nanolayer coatings
  for tool materials, *24*:426
Nanomaterials
  from sol–gel technology, *22*:526
Naphtha
  specific gravity, *23*:625
Naphthalene [*91-20-3*]
  in ethylene, *23*:456
  microbiological oxidation, *21*:609
  rejection by reverse osmosis membrane,
    21:319

sampling standards for, *21*:627
  sulfonation of, *23*:158
1,5-Naphthalene diisocyanate [*3173-72-6*],
    24:706
Naphthalenesodium, *22*:331
Naphthalene sulfonates, *23*:495
  in surfactants, *23*:158
1-Naphthalenesulfonic acid [*85-47-2*],
    23:194
2-Naphthalenesulfonic acid [*120-18-3*],
    23:194
Naphthalene
  boiling point of, *23*:627
Nardil, *22*:940
Naringin [*10236-47-2*], *23*:571
National Ambient Air Quality Standard
    (NAAQS), *21*:166; *22*:530
National Childhood Vaccine Injury Act of
    1986, *24*:745
National Emission Standards for
    Hazardous Air Pollutants (NESHAP),
    21:166
National Environmental Policy Act
    (NEPA), *21*:166
National Institute for Occupational Safety
    and Health, *21*:166
National Institute of Standards and
    Technology, *24*:627
National Pollutant Discharge Elimination
    System (NPDES), *21*:166
National Response Center (NRC), *21*:166
National Vaccine Injury Compensation
    Program, *24*:745
Natisite, *24*:264
Native asphalt, *23*:719
Natroautunite [*161334-19-6*], *24*:643
Natrolite
  silylating agents for, *22*:147
Natrosilite [*56941-93-6*], *22*:5
Natural gas
  fuel for steam prdn, *22*:746
  role in steelmaking, *22*:768
  sulfur removal from, *23*:250
  use in solar energy, *22*:472
Natural polymers
  as release agent, *21*:210
Natural resins, *21*:291, 297
Natural rubber [*9006-04-6*], *21*:483, 562
Naturals, *22*:323
Natural tannins
  as bird repellents, *21*:254
Naval stores, *21*:292

Navier-Stokes equations, *21*:668

Navigation aids
  solar energy for, *22*:475

Navigational devices
  thermoelectric power supplies for,
    23:1035

Navy vessels
  steam turbines for, *22*:755

Naxel
  alkylbenzensulfonate surfactant, *23*:496

Near-infrared absorption
  of triarylmethane dyes, *24*:567

Neat soap, *22*:303

Needles
  surgical for sutures, *23*:551

Nelles process, *24*:285

Nematodes
  veterinary drugs against, *24*:830

Neoabietic acid [*471-77-2*], *21*:294

Neodol
  alcohol surfactant, *23*:502
  ethoxylated alcohol surfactant, *23*:508

Neodox
  surfactant, *23*:493

Neohesperidin [*13241-33-3*], *23*:571

Neohesperidin dihydrochalcone [*20702-77-6*], *23*:571

Neomenthol [*491-01-0*], *23*:835

Neomycin [*119-04-0*], *24*:828

Neon
  diffusion in vitreous silica, *21*:1048

Neoprene
  as roofing material, *21*:446
  thiols in, *24*:28

Neosugar, *23*:25, 74

NEPA. See *National Environmental Policy Act.*

Nephila clavipes, *22*:155

Neral [*106-26-3*], *23*:835

Nernst glower
  for optical spectroscopy, *22*:635

Nerol [*106-25-2*], *23*:835

Nerolidol [*40716-66-3*], *23*:835, 857

Nerolidol alcohol [*7212-44-4*], *23*:870

Nerve agents
  detection of, *24*:509

Neryl acetate [*141-12-8*], *23*:855

NESHAP. See *National Emission Standards for Hazardous Air Pollutants.*

Network formation
  in silicones, *22*:94

Neuroregulators
  studies using stimulants, *22*:933

Neutralization
  of fatty acids, *22*:310

Neutron activation
  for selenium radioisotopes, *21*:686

Neutron activation analysis
  for trace analysis, *24*:498

Neutron diffraction, *21*:345

Neutron scattering
  for sol−gel technology studies, *22*:507

New Source Performance Standard
    (NSPS), *21*:166

Newsprint deinking mills, *21*:12

Newtonian limiting viscosity, *21*:350

Nextel, *21*:125

NFETs, *21*:742. (See also *nMOSFETs.*)

N/Furic, *22*:459

Nialamide [*51-12-7*], *22*:940

Niax, *24*:700

Nichrome
  for optical spectroscopy, *22*:635

Nickel
  adhesion in tire cord, *24*:171
  in brazing filler metals, *22*:492
  diffusion coefficient in vitreous silica,
    21:1047
  in ferrous shape-memory alloys, *21*:965
  plating with methanesulfonic acid,
    23:326
  role in steelmaking, *22*:780
  separation by reverse osmosis, *21*:322
  sulfuric acid as by-product, *23*:381
  tellurium in ores, *23*:782
  thin films of, *23*:1060

Nickel alloys
  as brazing filler metals, *22*:494
  for centrifuges, *21*:848
  as shape-memory alloys, *21*:964

Nickel carbonyl
  for chemical vapor deposition, *23*:1060

Nickel−chrome−boron
  for centrifuges, *21*:848

Nickel oxide
  solubility in steam, *22*:728

Nickel−rhenium alloys, *21*:340

Nickel−selenium [*1314-05-2*], *21*:698

Nickel steels
  for tanks and pressure vessels, *23*:644

Nickel sulfamate
  use in electroplating, *23*:130

Nickel titanate yellow, *24*:248

Niclosamide [50-65-7], 24:832
Nicotine
  detection of, 24:500
Nicotinic acid
  selenium in synthesis of, 21:713
Nigre soap, 22:303
NIKE
  vitreous silica optics for, 21:1066
Nimbus III, 23:1036
Ninol
  diethanolamine surfactant, 23:520
Niobium
  refractory coatings for, 21:105
  role in steelmaking, 22:779
  in shape-memory alloys, 21:968
  in steels, 22:811
  thorium oxides, 24:76
  tin alloys of, 24:120
  in titanium pigments, 24:248
NIST 14, 23:456
NIST Standard Reference Material 739,
  21:1052
Niter cake, 22:403
Niterox, 22:390
Nitinol, 21:964
Nitrates
  of poly(vinyl alcohol), 24:989
Nitration
  of sugar alcohols, 23:101
  of toluene, 24:357
Nitric acid [7697-37-2], 22:394
  boiling point of, 23:627
  sodium nitrate from, 22:388
  specific gravity, 23:625
  steel for manufacture of, 22:822
Nitric oxide [10102-43-9], 22:394, 654
Nitride fibers, 21:117
Nitrides
  as refractory coatings, 21:102
  in steel, 22:794
Nitriding
  of steels, 22:808
Nitrile
  thiols in, 24:28
Nitrile alloy membranes
  as roofing material, 21:448
Nitrile rubber, 21:487
  thiols in, 24:29
Nitrile silicone [70775-91-6], 22:110
Nitrilotriacetic acid
  water additive for steam prdn, 22:741

Nitrites
  determination of, 23:130
  in natural waters, 24:510
Nitrobenzene [98-95-3]
  as HAP compound, 22:532
  regulatory level, 21:160
  as solvent, 22:546
m-Nitrobenzenesulfonic acid [98-47-5],
  23:203
Nitrofurans
  as veterinary drugs, 24:827
Nitrogen
  diffusion in vitreous silica, 21:1048
  removal from natural gas, 23:139
  role in steelmaking, 22:779
  tanks for, 23:644
Nitrogen dioxide [10102-44-0], 22:394
  for dimethyl sulfoxide prprn, 23:225
Nitrogen oxides
  absorption of, 22:417
  sampling, 21:636
  sensors for, 21:824
Nitrogen selenide [12033-88-4], 21:700
Nitromethane [75-52-5]
  as solvent, 22:546
Nitronium tetrafluoroborate [13826-86-3],
  24:856
Nitrophenol
  rejection by reverse osmosis membrane,
  21:319
4-Nitrophenyl-β,β-dichloroethyl ketone
  [31689-13-1], 24:856
1-Nitropropane [105-03-2]
  as solvent, 22:546
2-Nitropropane [79-46-9]
  as HAP compound, 22:532
  as solvent, 22:546
5-Nitrosalicylic acid [96-97-9], 21:605
Nitrosamine regulation
  in the rubber industry, 22:1012
Nitrosamines
  from frankfurters, 24:515
N-Nitrosodiphenylamine [156-10-5],
  21:474; 22:400
Nitrosylsulfuric acid, 23:381
2-Nitrothiophene [609-40-9], 24:39
m-Nitrotoluene [99-08-1]
  as solvent, 22:546
4-Nitrotoluene-2-sulfonic acid [121-03-9]
  stilbene dye precursor, 22:922
Nitrous acid [7782-77-6], 22:394
Nitrous oxide [10024-97-2]
  supercritical fluid, 23:454

Nitryl oxide [14522-82-8], 24:835
NMA grouts
    ban on, 22:455
nMOSFETs, 21:742
Nmr. See Nuclear magnetic resonance.
Noise
    in centrifuge operation, 21:847
1-Nonanethiol [1455-21-6], 24:21
Nondestructive evaluation
    of refractory coatings, 21:113
Nonferrous metals
    impurity in selenium, 21:705
    use of sulfur for, 23:252
Nongrain-raising stains, 22:694
Nonisol
    polyoxyethylene surfactant, 23:514
Nonlinear optical spectroscopies, 22:657
Non-Newtonian fluids, 21:350
Nonoxide fibers, 21:117
Nonsteroidal antiinflammatory drug,
    24:502
Nonutility generators (NUGs), 21:181
Nonvolatile memories (NVMs), 21:745
Nonwoven fabrics
    vinyl acetate copolymers as binding
        agents for, 24:974
Nonylphenol ethoxylate
    sulfation of, 23:172
Nopalcol
    polyoxyethylene surfactant, 23:514
Noradrenaline
    trace analysis of, 24:507
Nordel
    ethylene–propylene polymer, 21:486
Norepinephrine [51-41-2], 22:937
Norethindrone [68-22-4], 22:880, 903
Norethynodrel [68-23-5], 22:880
Norgestimate [35189-28-7], 22:903
Norgestrel [6533-00-2], 22:903
D-Norgestrel [797-63-7], 22:892
Normal sugar solution, 23:13
(17α)-19-Norpregna-1,3,5(10)-trien-20-yne-
    3,17-diol, 22:903
Nortriptyline [72-69-5], 22:942
Notice of Proposed Rulemaking (NPRM),
    21:166
Nougat
    vanillin in, 24:820
Novacekite [12255-29-1], 24:643
Novafil, 23:543
NovaFlex process, 24:711
NOVA laser
    vitreous silica optics for, 21:1065

No-wipe stains, 22:694
NO_x control
    in power generation, 21:188
NOXSO, 22:346
Nozzle inserts
    refractory coatings for, 21:106
NPDES. See National Pollutant Discharge
    Elimination System.
NPRM. See Notice of Proposed
    Rulemaking.
NRC. See National Response Center;
    Nuclear Regulatory Commission.
NSG-ES, 21:1034
NSPS. See New Source Performance
    Standard.
Nuclear fuel
    thorium-232 as, 24:71
    thorium oxalate in, 24:78
Nuclear magnetic resonance (nmr),
    22:505
    ^{13}C for silk structure defn, 22:157
    silicon-29 for hydrolysis studies, 22:505
Nuclear power
    steam cycle, 22:752
Nuclear quadrupole resonance
    of potassium chlorate, 23:828
Nuclear reactors
    thorium for, 24:72
    use of sodium in, 22:345
Nuclear Regulatory Commission (NRC),
    21:166, 190
Nuclear waste
    from power generation, 21:191
    trace analysis of, 24:518
Nucleosides
    silylation in synthesis of, 22:146
NUGs. See Nonutility generators.
Nukiyama-Tanasawa equations, 22:678
Nurolon, 23:543
n-Nutane [106-97-8]
    supercritical fluid, 23:454
Nutmeg
    extraction using SCFs, 23:466
NutraSweet, 23:558
Nutrition
    of soybeans and other oilseeds, 22:608
Nuts
    soybean substitutes, 22:614
NVM. See Nonvolatile memories.
Nylon
    mechanical properties of, 22:160
    silane coupling agent for, 22:152

in tire cord, *24*:163
tire cords, *24*:162
Nylon-6 [*25038-54-4*]
  as sutures, *23*:542
  sutures from, *23*:550
  in tire cord, *24*:163
Nylon-6,6 [*32131-17-2*]
  critical surface tension of, *22*:148
  for sutures, *23*:550
  in tire cord, *24*:163
Nylon trogamid
  sol–gel–silica hybrids of, *22*:526
Nystatin [*34786-70-4*], *24*:829

O

OBTS. See *N-Oxydiethylene-2-benzothiazolesulfenamide.*
Occupational Safety and Health Act,
  21:166
Ocean
  source of sulfur, *23*:246
Octadecamethyloctasiloxane [*556-69-4*],
  22:103
Octadecanethiol
  monolayers, *23*:1096
1-Octadecanethiol [*2885-00-9*], *24*:21
Octadecanoic acid
  as rubber chemical, *21*:473
Octadecyl acrylate
  thin films of, *23*:1081
Octadecylamine
  water additive of steam prdn, *22*:740
Octadecylamine monolayer [*124-30-1*]
  release substrate, *21*:212
4-Octadecylamino-4'-nitroazobenzene
  thin films of, *23*:1085
n-Octadecyl hydrogen sulfate [*143-03-3*],
  23:410
Octadecyltrichlorosilane [*112-04-9*], *22*:149
Octadecyltriethoxysilane [*7399-00-0*],
  22:72
Octadecyltrimethoxysilane [*3069-42-9*],
  22:72
  self-assembled monolayers of, *23*:1090
Octadecyltrimethylammonium bromide,
  21:1019
Octafluoro-1*H*-pentanoic acid, *24*:300
Octamethylcyclotetrasiloxane [*556-67-2*],
  22:103
Octamethyltrisiloxane [*107-51-7*], *22*:103

n-Octane [*111-65-9*]
  as solvent, *22*:536
1-Octanethiol [*111-88-6*], *24*:21
2-Octanol [*123-96-6*]
  as solvent, *22*:536
Octyltriethoxysilane [*2943-75-1*], *22*:72
Odorant
  thiols for natural gas, *24*:30
Off!
  insect repellent, *21*:243
Off-gases
  sulfur from, *23*:250
Ofner method
  reducing sugar test, *23*:16
Ohmic contacts, *21*:736, 805
Oil. See *Fats and oils*; *Petroleum.*
Oil and gas drilling
  sodium bromide for, *22*:379
Oilfield fluids
  organic titanates as, *24*:332
Oil laundering, *21*:1
Oil mining, *23*:730
Oil of bergamot
  in tea, *23*:759
Oil of Olay, *22*:323
Oil of turpentine
  boiling point of, *23*:627
Oil reclaiming, *21*:1
Oil recovery
  centrifugal separation for, *21*:848
Oil recycling, *21*:1
Oil reprocessing, *21*:2
Oils
  hydrogels in refining of, *21*:1001
  nir spectroscopy of, *22*:641
  removal using SCFs, *23*:466
  soap source, *22*:297
  soybean and other oilseeds, *22*:610
  sulfation of, *23*:171
Oils, sulfonated, *23*:479
Oil slicks, *23*:113
  dispersion, *23*:212
Olefins
  catalyst for stereospecific
    polymerization, *24*:257
  polymerization, *23*:209
  silver complexes of, *22*:185
  sulfurization of, *23*:429
α-Olefins
  fractionation using SCFs, *23*:469
  sulfonation of, *23*:160
α-Olefin sulfonate, *23*:159

Olefin sulfonates, 23:497
Oleic acid [112-80-1], 23:617
    as release agent, 21:210
    separation coefficients for, 21:926
Oleochemicals
    as surfactants, 23:479
Oleoresin capsicum [8023-77-6], 21:259
Oleoresins, 23:833
Olestra, 23:7, 68
Oleum [8014-95-7], 23:195, 373
    for sulfonation, 23:147
Oleyl palmitamide [16260-09-6]
    as release agent, 21:210
Olibanum [8050-07-5], 21:299
Oligosiloxanes, 22:91
Olive oil
    specific gravity, 23:625
Olives, 23:96
    sorbates in, 22:583
Olive tree
    mannitol in, 23:97
Olivine, 21:980
    magnetic intensity, 21:876
    silylating agents for, 22:147
One-shot method
    silicate grouting, 22:453
Onions
    mannitol in, 23:97
Oolong tea, 23:753
Oospora sp.
    inhibited by sorbates, 22:580
Opacity
    of titanium dioxide, 24:240
Open-pit mining, 23:730
Ophthalmia neonatorum
    silver nitrate for, 22:191
Opisthotonos, 22:933
Opposed jet mills, 22:293
Optical amplifiers
    compound semiconductors, 21:788
Optical coating technology
    thin films for, 23:1046
Optical counters
    use in particle size measurement,
        22:276
Optical fibers
    coating of, 23:1074
    compound semiconductors, 21:788
    sensors based on, 21:824
    of vitreous silica, 21:1065
Optical lithography
    compound semiconductors, 21:809
    optics of synthetic fused silica, 21:1063

Optical multichannel analyzers, 22:643
Optical rotation
    of sugar solutions, 23:13
Optical rotatory dispersion, 22:651
Optical waveguides
    vitreous silica in, 21:1000
Optoelectronic imaging devices, 22:643
Optosil, 21:1034
Oral polio virus vaccine, 24:728
Oral toxicity
    for sugar alcohols, 23:106
Orange juice
    trace analysis of, 24:515
Orange pekoe, 23:758
Orcinol, 24:79
Ore
    transportation of, 24:524
Ore flotation
    sodium sulfite used in, 23:314
Ore processing
    use of sulfuric acid, 23:397
Organic acid
    for fluxes, 22:496
Organic esters
    from poly(vinyl alcohol), 24:990
Organic felts
    as roofing material, 21:442
Organic grouting systems, 22:454
Organic–inorganic hybrids
    from sol–gel technology, 22:525
Organic tire cord, 24:164
Organogels, 23:113
Organometallic chemistry
    of thorium, 24:73
Organophosphates
    degradation in soil, 22:430
Organophosphonates
    sensors for, 21:824
    water additive of steam prdn, 22:740
Organosilanes, 22:69
    from silanes, 22:47
Organosilicates
    silica from, 21:1019
Organosilicone coatings, 22:116
Organosphosphates
    as veterinary drugs, 24:830
Organotin compounds, 24:131
Orifice viscometers, 21:379
Orkla process, 23:245
Orleans process, 24:843
Ornitrol, 21:256

Orthodontics
  brazing filler metals for, *22*:490
  shape-memory alloys for, *21*:973
Orthopedic devices
  shape-memory alloys for, *21*:973
Orthosilicic acid, *21*:1010; *22*:69
Orthotelluric acid [*7803-68-1*], *23*:796, 798
Orthotitanic acid [*20338-08-3*], *24*:234
Oscillators
  synthetic quartz crystals for, *21*:1082
Osladin [*33650-66-7*]
  potential sweetener, *23*:573
Osmotic pressure
  role in reverse osmosis, *21*:307
Osram process, *21*:1038
Osteoporosis drugs
  steroids as, *22*:852
Ostwald glass capillary viscometer, *21*:376
Ostwald ripening
  role in sol–gel technology, *22*:509
  of silica, *21*:1016
OTC. See *Over-the-counter*.
Otitis media
  vaccine against, *24*:737
OTOS. See *N-Oxydiethylenethiocarbamyl-N-oxydiethylenesulfenamide*.
Outdoor fabrics, *23*:907
Outokumpu process, *23*:245
Over-the-counter (OTC) drugs, *21*:173
Overtone stains, *22*:696
Oxaloacetic acid [*328-42-7*]
  from succinic acid, *22*:1077
1,4-Oxatellurane [*5974-87-8*], *23*:800
Oxendolone [*33765-68-3*], *22*:882
Oxfendazole [*53716-50-0*], *24*:831
Oxibendazole [*20559-55-1*], *24*:831
Oxidants
  in organic synthesis, *23*:959
Oxidative bleaching
  in paper recycling, *21*:18
Oxide fibers, *21*:117
Oxidized starches, *22*:713
O_2-Oxidoreductase, *23*:751
Oxo ion salts
  in reprocessing uranium, *24*:665
11-Oxoprogesterone [*516-15-4*], *22*:882
Oxtriphylline, *22*:936
p,p'-Oxybis(benzenesulfonyl hydrazide) [*80-51-3*]
  as blowing agent, *21*:479
2,2'-Oxybisethanethiol [*2150-02-9*], *24*:21
Oxychozanide [*2277-92-1*], *24*:832

Oxydemeton-methyl
  thiols in, *24*:29
Oxydeprofos
  thiols in, *24*:29
N-Oxydiethylene-2-benzothiazole-sulfenamide (OBTS) [*102-77-2*], *21*:465
N-Oxydiethylenethiocarbamyl-*N*-oxydiethylenesulfenamide (OTOS) [*13752-51-7*], *21*:465
Oxygen [*7782-44-7*], *24*:854
  atmospheric, analysis using Raman spectroscopy, *22*:650
  atomic in silver, *22*:168
  in brazing filler metals, *22*:492
  diffusion in vitreous silica, *21*:1048
  microsensor for, *21*:824
  in paper recycling, *21*:18
  solubility in sodium, *22*:330
  as steam component, *22*:721
  in steelmaking, *22*:773
  at sulfuric acid plants, *23*:395
  tanks for, *23*:644
  use in optical spectroscopy, *22*:645
  water additive of steam prdn, *22*:740
Oxygenated solvents, *22*:535
Oxygen radical absorbance capacity
  of tea extracts, *23*:763
Oxygen scavenger
  in steam, *22*:737
Oxytetracycline [*79-57-2*], *24*:829
  as veterinary drugs, *24*:827
Oxytocin [*50-56-6*], *24*:833
Ozone [*10028-15-6*], *22*:654; *24*:855
  control in power generation, *21*:189
  in paper recycling, *21*:18
  in textile finishing, *23*:910
Ozone depletion potential
  of refrigerants, *21*:133
Ozone pollution, *22*:532

P

P-4000 [*553-79-7*]
  potential sweetener, *23*:573
Pacemakers
  solid tantalum capacitors in, *23*:670
Packaging
  for sterilized materials, *22*:846
  for sutures, *23*:552

of tea, *23*:760
use in sterile filling, *22*:846
Packaging materials
    titanate coatings, *24*:324
Pad stains, *22*:697
Paget's disease, *24*:101
Paints
    ethoxysilanes in, *22*:79
    nonemulsion, *24*:323
    prepared from poly(vinyl acetate),
        24:972
    of PVAc, *24*:972
    PVC emulsions for, *24*:1030
    rheological measurements, *21*:427
    sampling standards for, *21*:627
    solvents for, *22*:568
    talc application, *23*:613
    titanated thixotropics, *24*:327
    titanium dioxide in, *24*:240
    use of sulfur for, *23*:251
Palatinit, *23*:25, 74
Palatinose [*13718-94-0*], *23*:8, 25, 74
Pale crepes, *21*:565
Palladium [*7440-05-3*], *21*:107; *24*:852
    in brazing filler metals, *22*:495
    in ferrous shape-memory alloys, *21*:965
Palmitic acid [*57-10-3*], *23*:617
    as release agent, *21*:210
    separation coefficients for, *21*:926
Palmitine, *24*:506
Palm kernel oil [*8023-79-8*], *22*:302
Palm oil [*8002-75-3*], *22*:302
    specific gravity, *23*:625
Palmolive, *22*:323
Pal Sweet, *23*:559
Palustric acid [*1945-53-5*], *21*:294
Panqueque, *22*:383
Papagoite [*12355-62-3*], *22*:3
Paper, *24*:13
    poly(vinyl acetate) as binder, *24*:973
    reverse osmosis for wastewater, *21*:322
    sampling in mills, *21*:641
    sampling standards for, *21*:627
    sedimentation in, *21*:682
    silica coatings for, *21*:1020
    sizing of, *24*:329
    steel for manufacture of, *22*:823
    sulfonic acid-derived dyes in, *23*:206
    talc application, *23*:613
    for tea bags, *23*:759
    triarylmethane dyes for, *24*:565
    use of sodium sulfate, *22*:410

use of sulfur, *23*:251
use of sulfur dioxide, *23*:310
use of sulfuric acid, *23*:397
use of ultrafiltration, *24*:622
Paperboard
    sampling standards for, *21*:627
Paper chromatography
    of sugar alcohols, *23*:103
Paper colorants
    stilbene dyes as, *22*:927
Paper electrophoresis
    of polyols and carbohydrates, *23*:103
Papermaking
    starch in, *22*:712
    titanium dioxide for, *24*:249
Paper mills
    scrap tire fuel, *21*:25
    steam system for, *22*:754
Paper recycling mills, *21*:10
Paper release coatings
    silicone products, *22*:116
Paper sizes
    PVA resins as, *24*:980
    rosin as, *21*:296
PAPI, *24*:697
Paprika
    extraction using SCFs, *23*:466
Papularia arundinis
    inhibited by sorbates, *22*:579
Parabens, *21*:602
Paraffin wax [*8002-74-2*]
    critical surface tension of, *22*:148
    as release agent, *21*:210
    release substrate, *21*:212
Parainfluenza
    vaccine against, *24*:737
Parallel plate viscometers, *21*:391
Pararosaniline [*569-61-9*], *24*:552
Parasiticides
    as veterinary drugs, *24*:830
Parasorbic acid [*108-54-3*], *22*:577
Parathion [*56-38-2*], *23*:291
Parazylene, *21*:864
Parnate, *22*:940
Paroxetine [*61869-08-7*], *22*:944
Parsonsite [*12137-57-4*], *24*:643
Particle bonding, *22*:223
Particle breakage, *22*:279
Particle-induced x-ray emission, *22*:655
Particle size enlargement, *22*:222
Particulate deposition
    for refractory coatings, *21*:98

Parting agents, 21:207
Partition chromatography
   of sugar alcohols, 23:103
Passenger tire equivalents (PTE), 21:23
Passenger tires, 24:172
Passive immunization, 24:737
Passive solar devices, 21:225
Pasteurella sp.
   antimicrobial agent for, 24:828
Pasteurization, 22:847
Pasteurizers
   sulfamic acid cleaner for, 23:129
Patchouli alcohol [5986-55-0], 23:872
Patent Blue V [3546-49-0], 24:556
Patronite [12188-60-2], 24:783
Pauling's rules
   for silica, 21:980
Paving materials
   sulfur in, 23:261
Paxil, 22:945
PCBs. See *Polychlorinated biphenyls.*
PDS II, 23:543
Peaches, 23:96
Peanut butter, 22:615
   dextrose in, 23:592
   sorbitol in, 23:110
Peanuts, 22:591
   extraction using SCFs, 23:467
Pearlite, 22:790
   phase properties of, 22:797
Pearl polymerization, 24:958
Pears, 23:96
Peat
   sampling standards for, 21:627
PEB. See *Polyethylbenzenes.*
Pebulate
   thiols in, 24:30
PEC-1000, 21:318
PECVD. See *Plasma-enhanced chemical
   vapor deposition.*
Pegmatites
   tantalite in, 23:660
Pegosperse
   polyoxyethylene surfactant, 23:514
Pegosperse 50 DS
   fatty acid surfactant, 23:518
Pegosperse 50 MS
   fatty acid surfactant, 23:518
Pegosperse 100 O
   fatty acid surfactant, 23:518
Pegosperse 100 S
   fatty acid surfactant, 23:518

Pekoe fannings, 23:758
Pelletizing, 22:242
Pellet mills, 22:239
Peltier effect, 23:805, 1030, 1031
Pemoline [2152-34-3], 22:937
Pendimethalin [40487-42-1], 22:422
Pendulum mills, 22:287, 288
Penetrating stains, 22:694
Penetrometers, 21:409
Penicillin amidase
   deactivation of, 23:467
Penicillins
   silylation in synthesis of, 22:146
Penicillium atromentosum
   inhibited by sorbates, 22:579
Penicillium chermesinum
   inhibited by sorbates, 22:579
Penicillium chrysogenum
   inhibited by sorbates, 22:579
Penicillium citrinum
   inhibited by sorbates, 22:579
Penicillium digitatum
   inhibited by sorbates, 22:579
Penicillium duclauxi
   inhibited by sorbates, 22:579
Penicillium expansum
   inhibited by sorbates, 22:579
Penicillium frequentans
   inhibited by sorbates, 22:579
Penicillium funiculosum
   inhibited by sorbates, 22:579
Penicillium gladioli
   inhibited by sorbates, 22:579
Penicillium herquei
   inhibited by sorbates, 22:579
Penicillium implicatum
   inhibited by sorbates, 22:579
Penicillium italicum
   inhibited by sorbates, 22:579
Penicillium janthinellum
   inhibited by sorbates, 22:579
Penicillium notatum
   inhibited by sorbates, 22:579
Penicillium oxalicum
   inhibited by sorbates, 22:579
Penicillium patulum
   inhibited by sorbates, 22:579
Penicillium piscarium
   inhibited by sorbates, 22:579
Penicillium purpurogenum
   inhibited by sorbates, 22:579
Penicillium restrictum
   inhibited by sorbates, 22:579

Penicillium roquefortii
  inhibited by sorbates, 22:579
Penicillium rugulosum
  inhibited by sorbates, 22:579
Penicillium spp.
  inhibited by sorbates, 22:579
Penicillium sublateritium
  inhibited by sorbates, 22:579
Penicillium thomii
  inhibited by sorbates, 22:579
Penicillium urticae
  inhibited by sorbates, 22:579
Penicillium variabile
  inhibited by sorbates, 22:579
Penning discharge, 22:43
2,3,4,3′,4′-Penta-O-acetylsucrose [34382-02-2], 23:66
2,3,6,1′,6′-Penta-O-acetylsucrose [35867-25-5], 23:66
Pentacarbonato thorium(IV) [12364-90-8], 24:77
Pentachlorophenol [87-86-5]
  regulatory level, 21:160
  rejection by reverse osmosis membrane, 21:319
Pentachlorothiophenol [133-49-3]
  in rubber processing, 21:478
Pentadin [61391-05-7]
  potential sweetener, 23:573
Pentaerythritol, 24:290
  use in x-ray spectroscopy, 22:654
Pentaerythritol disulfite [3670-93-7], 23:410
Pentaerythritol-tetrakis-3-mercapto-propionate [7575-23-7], 24:7
Pentaerythritol-tetrakis-thioglycolate [10193-99-4], 24:7
Pentamethyldiethylenetriamine, 24:699
Pentamethyldipropylenetriamine, 24:699
Pentane [109-66-0]
  as solvent, 22:536
n-Pentane [109-66-0]
  dielectric constants of, 22:759
  supercritical fluid, 23:454
Pentanesulfonic acid [35452-30-3], 23:194
1-Pentanethiol [110-66-7], 24:21
1-Pentanol [71-41-0]
  as solvent, 22:536
Pentflic acid, 23:209
Pentlandite [53809-86-2]
  sulfide ore, 23:244
  tellurium in, 23:782

Pentobarbital [57-33-0], 24:835
Pentylenetetrazol [54-95-5], 22:934
Pentyl propionate [624-54-4]
  as solvent, 22:542
Pentyltriethoxysilane [2761-24-2], 22:72
Pepper spray, 21:259
Peptides, 24:506
Peptide vaccine, 24:742
Peptization
  role in sol–gel technology, 22:521
Perchloroethylene [127-18-4]
  as HAP compound, 22:532
  as solvent, 22:538
Percolation theory, 22:507
Perdiselenic acid [81256-78-2], 21:701
Perfluorolauric acid [307-55-1]
  as release agent, 21:210
Perfluorolauric acid monolayer
  release substrate, 21:212
Perfumery
  vanillin in, 24:822
Perhydro-1,2-cyclopentenophenanthrene
  ring system, 22:851
Periclase [1309-48-4], 21:55
Perillartine [30950-27-7]
  potential sweetener, 23:573
Permaflex, 24:1079
Permalloy
  thin films of, 23:1071
Permanent press, 23:900
Permanone Tick Repellent, 21:250
Permeability coefficient
  role in reverse osmosis, 21:309
Permissible exposure limit, 21:166
Permonoselenic acid [81256-77-1], 21:701
Perovskite [12194-71-7], 24:230
Peroxidase
  in teas, 23:751
Peroxides
  cross-linking agent, 21:471
  titanium complexes, 24:299
Peroxymonosulfuric acid [7722-86-3], 23:312
Peroxysuccinic acid [2279-96-1], 22:1077
Perrhenic acid [13768-11-1], 21:338
Personal cleansing products, 22:324
Personal computers
  solid tantalum capacitors in, 23:670
Pertussis vaccine, 24:731
Pesticide residues
  in soil, 22:419
Pesticides, 23:113, 466
  analysis using fluorescence, 22:653

carrier for, *21*:1025
  sampling standards for, *21*:627
  solvents for, *22*:570
  use of sulfur for, *23*:252
Pestolotiopsis macrotricha sp.
  inhibited by sorbates, *22*:579
Pet foods
  sugar alcohols in, *23*:109
  use of sorbates, *22*:584
Petro
  naphthalene sulfonate surfactant,
    23:497
Petrochemical
  reverse osmosis for wastewater, *21*:322
Petrochemistry
  use of tantalum fluoride, *23*:676
Petroleum, *22*:250, 422; *23*:718
  centrifugal separation for, *21*:848
  enhanced recovery by steam injection,
    22:758
  measurement of particles in, *22*:276
  recycling of, *21*:1
  refractories for the industry, *21*:83
  rhenium catalysts for, *21*:342
  sampling standards for, *21*:627
  specific gravity, *23*:625
  steel quenching in, *22*:801
  steels for the industry, *22*:818
  sulfur removal from, *23*:250
  sweetening of, *23*:138
  tanks for, *23*:647
  transportation of, *24*:524
Petroleum additives
  thiols for, *24*:30
Petroleum alkylation
  use of sulfuric acid, *23*:397
Petroleum plants
  shape-memory alloys for, *21*:969
Petroleum refining
  thioglycolic acid for, *24*:16
  use of sulfur for, *23*:252
Petroleum re-refining, *21*:2
Petroleum sulfonates, *23*:162
  in enhanced oil recovery technology,
    23:212
Petroleum waxes
  as release agent, *21*:210
Petrosil, *21*:1058
Petrov catalysts, *23*:211
Petzetakis, *24*:710
Petzite [*1317-73-3*]
  tellurium in, *23*:783

Pewter
  tin alloy, *24*:120
PFETs, *21*:742. (See also *pMOSFETs*.)
P&G Amide No. 27
  surfactant, *23*:520
pH Adjustment
  sulfamic acid for, *23*:129
Pharmaceutical applications
  of vanillin, *24*:823
Pharmaceuticals
  analysis using fluorescence, *22*:653
  analysis using infrared spectroscopy,
    22:641
  analysis using Raman spectroscopy,
    22:651
  analysis using vibrational circular
    dichroism, *22*:651
  dextrose in prdn of, *23*:592
  optical spectroscopic analysis, *22*:644
  regulation agencies, *21*:172
  rubidium compounds in, *21*:598
  saccharin in, *23*:566
  SCF solution of, *23*:472
  silylation in synthesis of, *22*:146
  sodium nitrite for, *22*:394
  thiophene and thiophene derivatives as,
    24:47
  triarylmethane dyes for, *24*:565
  use of hexitols in, *23*:107
  use of sulfur for, *23*:251
  from vanillin, *24*:823
  as veterinary drugs, *24*:826
Phase diagrams
  of refractories, *21*:58
Phase Doppler particle analyzers, *22*:686
Phase rule, *23*:811
Phase separation
  SCFs induction of, *23*:469
Phase structure
  of soap, *22*:302
Phase-transfer catalysts
  sulfonic acids as, *23*:206
PHD polyols. See *Polyharnstoff dispersion
  polyols*.
(−)-β-Phellandrene [*6153-17-9*], *23*:860
Phenanthrene
  rejection by reverse osmosis membrane,
    21:319
o-Phenanthroline ferrous–sulfate
  for tellurium analysis, *23*:792
Phenelzine [*51-71-8*], *22*:939
Phenethyl salicylate [*87-22-9*], *21*:615

Phenobarbital [50-06-6], 24:835
Phenol [108-95-2], 24:856
    rejection by reverse osmosis membrane, 21:319
    sampling standards for, 21:627
Phenol cake, 22:403
Phenol–formaldehyde resins
    in grouting systems, 22:457
Phenolic resins
    silane coupling agent for, 22:152
    use in refractory brick, 21:69
Phenolphthalein [77-09-8], 24:567
Phenol red [143-74-8], 24:567
Phenols, 23:466
    reactions with silanes, 22:51
Phenoplasts
    in grouting systems, 22:457
Phenosafranine, 22:190
Phenothiazine [58-37-7], 24:831
Phenothiol
    thiols in, 24:29
Phentermine [122-09-8], 22:938
Phenylalanine
    in silk, 22:156
    in soybeans and other oilseeds, 22:596
Phenylalanine ammonia lyase, 23:751, 752
Phenylaminoethyl selenide, 21:713
5-Phenylazosalicylic acid [3147-53-3], 21:607
Phenylbutazone [50-33-9], 24:832
3-Phenyl-1-butene [934-10-1], 24:854
4-Phenyl-1-butene [768-56-9], 24:854
N,N'-meta-Phenylenebismaleimide [3006-93-7]
    cross-linking agent, 21:471
p-Phenylene diisocyanate [104-49-4], 24:706
1-Phenylethylmagnesium bromide [41745-02-2], 24:853
2-Phenylethylmagnesium bromide [3277-89-2], 24:853
Phenyl hydrogen sulfate [937-34-8], 23:410
Phenylmethylsilane [766-08-5], 22:60
Phenylmethylsulfinylbenzene [833-82-9], 23:218
N-Phenyl-4[[4-(phenylamino)phenyl](4-(phenylimino)-2,5-cyclohexadien-1-ylidene)methyl]-benzeneamine, monohydrochloride [2152-64-9], 24:554

Phenylpropanolamine [14838-15-4], 22:938
Phenyl salicylate [118-55-8], 21:614
Phenyltellurinic acid [83270-41-1], 23:800
Phenyltriethoxysilane [780-69-8], 22:72, 147
Phenyltrimethoxysilane [2996-92-1], 22:153
    critical surface tension of, 22:148
Phenyl vinyl ether [766-94-9], 24:1056
Phlebitis
    shape-memory alloy device for, 21:973
Phloroglucinol, 24:79
Phoma sp.
    inhibited by sorbates, 22:579
Phorate
    thiols in, 24:29
Phosgene [75-44-5], 24:855
Phosphatation
    of sugar liquor, 23:35
Phosphate esters
    of sugar alcohols, 23:101
    of sugars, 23:69
Phosphate fertilizer
    tetrafluorosilane as by-product, 22:64
Phosphates
    of poly(vinyl alcohol), 24:989
Phosphate slimes
    thickening of, 21:680
Phosphatides
    in soybeans and other oilseeds, 22:596
Phosphine
    dopant for pure silicon, 21:1095
    reactions with silanes, 22:52
Phosphogypsum
    source of sulfur, 23:247
Phosphorescence, 22:652
Phosphoric acid
    centrifugal separation of, 21:860
    specific gravity, 23:625
    use in refractory brick, 21:70
    use of sulfuric acid, 23:397
    vanadium as by-product, 24:783
Phosphorimetry, 22:653
Phosphorous pentoxide
    use in silica sol–gel systems, 22:527
Phosphors
    particle size measurement for, 22:270
    rare-earth, 23:829
    silicates as binders, 22:23
Phosphorus
    in brazing filler metals, 22:484, 492
    dopant for silicon, 21:1098

in pig iron, *22*:767
in silicon and silica alloys, *21*:1106
in steel, *22*:775
vitreous silica impurity, *21*:1036
Phosphorus oxychloride [*10025-87-3*], *23*:195
Phosphorus pentabromide [*7789-69-7*], *23*:195
Phosphorus pentachloride [*10026-13-8*], *23*:195
Phosphorus pentoxide [*1314-56-3*], *23*:197
Photoacoustic spectroscopy, *22*:639
Photochemistry
    optical spectroscopy for, *22*:643
Photochromic glass
    silver in, *22*:175
Photoconductors
    compound semiconductors in, *21*:796
    for optical spectroscopy, *22*:635
Photocopiers
    use of selenium, *21*:709
Photodegradation
    of polystyrene, *22*:1033
Photodetectors
    as sensors, *21*:820
Photodiodes
    compound semiconductors in, *21*:796
    for optical spectroscopy, *22*:635
    titanium coatings, *24*:323
Photoextinction
    use in particle size measurement, *22*:269
Photographic plates
    use in atomic emission spectroscopy, *22*:646
Photography
    selenium cmpds in, *21*:714
    silver cmpds for, *22*:192
    sodium bromide for, *22*:379
    sodium sulfite used in, *23*:314
    sodium thiosulfate, *24*:60
Photoimaging systems
    triarylmethane dyes, *24*:565
Photoionization
    of atoms or molecules, *22*:657
Photolithography, *23*:1073
Photolysis
    of pesticides, *22*:434
    of silanes, *22*:43
    of titanium compounds, *24*:319
Photomicrolithography
    triarylmethane dyes, *24*:566

Photomultipliers
    use in optical spectroscopy, *22*:635, 643
Photon correlation spectroscopy
    for particle size measurement, *22*:270
Photon detectors
    compound semiconductors in, *21*:796
Photonics technology
    compound semiconductors, *21*:788
Photoresists, *23*:210
    thin films for, *23*:1081
Photosynthesis, *22*:476
Photothermal spectroscopy, *22*:659
Photovoltaic (PV) cells
    as renewable energy source, *21*:219
Photovoltaic devices
    compound semiconductors in, *21*:796
Photovoltaic photometric cell
    selenium, *21*:709
Photovoltaics
    pure silicon, *21*:1101
Photozone counters, *22*:276
Phthalic anhydride [*85-44-9*]
    as rubber chemical, *21*:473
    sampling standards for, *21*:627
(Phthalocyaninato(2-))copper, *23*:551
Phyllodulcin [*55555-33-4*]
    potential sweetener, *23*:573
Physical vapor deposition
    for thin films, *23*:1041
    for tool coatings, *24*:424
    use on steel tool materials, *24*:403
Phytic acid [*83-86-3*]
    in soybeans and other oilseeds, *22*:598
Phytol [*505-06-5*], *23*:873
Piano wire, *22*:825
Pichi
    sugar alcohols in, *23*:96
Pichia alcoholophila
    inhibited by sorbates, *22*:580
Pichia membranaefaciens
    inhibited by sorbates, *22*:580
Pichia polymorpha
    inhibited by sorbates, *22*:580
Pichia silvestris
    inhibited by sorbates, *22*:580
Pichia sp.
    inhibited by sorbates, *22*:580
Pickles
    sorbates in, *22*:583
Picric acid [*88-89-1*]
    from salicylic acid, *21*:605
Picrochromite, *21*:56

Picrotin [21416-53-5], 22:932
Picrotoxin [124-87-8], 22:932
Picrotoxinin [17617-45-7], 22:932
Pidgeon process, 21:1114
Piezoelectric atomizer, 22:673
Pig iron
    role in steelmaking, 22:767
Pigment Blue 61 [1324-76-1], 24:557
Pigmented wood stains, 22:694
Pigment lakes
    from triarylmethane dyes, 24:557
Pigments, 23:113
    analysis using Raman spectroscopy,
        22:650
    cadmium sulfoselenide, 21:712
    cadmium telluride, 23:804
    particle size measurement for, 22:270
    pretreated with titanates, 24:323
    in soap, 22:319
    strontium compounds as, 22:953
    sulfamic acid in mfg, 23:129
    tantalum nitrides as, 23:676
    thioglycolic acids in, 24:16
    titanium compounds as, 24:248
    titanium dioxide for, 24:239
    use of sulfur for, 23:251
Piloted airblast atomizer, 22:672
Pimaric acid [127-27-5], 21:294
Pinacols
    titanates of, 24:289
Pinacolylmethylphosphonic acid, 24:509
cis-Pinane [6876-13-7], 23:834
Pin disk mills, 22:290
Pinene
    sampling standards for, 21:627
α-Pinene [80-56-8], 23:834
    as solvent, 22:536
β-Pinene [127-91-3], 23:834
    as solvent, 22:536
(+)-cis-α-Pinene epoxide [1686-14-2],
        23:861
Pine oil [8002-09-3]
    sampling standards for, 21:627
    as solvent, 22:548
Pioloform F, 24:936
Pioneer
    thermoelectric generators for, 23:1035
Pipelines
    sampling in, 21:641
    solders for, 22:487
Pipeline transportation, 24:530
Piperidinium pentamethylene-
        dithiocarbamate [98-77-1], 21:466

(−)-cis-Piperitol [65733-28-0], 23:861
(+)-trans-Piperitol [65733-27-9], 23:861
Pipes
    PVC use, 24:1040
    sampling in, 21:639
    tin, 24:115
Piping
    role in tank leaks and spills, 23:651
Piston cylinder capillary viscometer,
        21:382
Pitch, 23:679, 719
    fractionation using SCFs, 23:469
    sampling standards for, 21:627
    use in refractory brick, 21:71
Pitchblende [1317-75-5], 24:638, 642, 648
Pitot tube
    use in sampling, 21:633
pK_a values
    for triazines, 22:43
Plane tree
    mannitol in, 23:97
Planiblock, 24:710
Plants
    selenium in, 21:689
Plant steroids, 22:852
Plasdone, 24:1094
Plasma
    analysis using laser-induced
        fluorescence, 22:653
    use in thin film formation, 23:1051
Plasma-enhanced chemical vapor
        deposition (PECVD), 23:1060
    amorphous silicon, 21:751
Plasma-induced polymerization
    organosilicon monomers, 22:94
Plasma technology
    for diamond films, 24:439
    etching of silicon, 21:1089
    use for refractory coatings, 21:96
    use in atomic emission spectroscopy,
        22:647
Plasmons, 23:1087
Plastic hardcoats
    silicone products, 22:116
Plasticity retention index
    of rubber, 21:571
Plasticization
    of PVAc, 24:968
Plasticizers, 21:508, 527
    organotin catalysts for, 24:145
    for PVB resins, 24:930
    sampling standards for, 21:627

sorbitol solutions as, *23*:108
in styrene plastics, *22*:1058
supercritical fluid as, *23*:460
titanate catalysts for, *24*:290
use of sulfolane, *23*:139
Plastics
    rheological measurements, *21*:427
    sampling standards for, *21*:627
    for size separation screens, *21*:908
    talc application, *23*:613
    triarylmethane dyes for, *24*:565
    use of sulfuric acid, *23*:397
Plastic working
    of steel, *22*:788
Plate glass
    use of selenium in, *21*:712
Platforming process, *24*:360
Plating
    thin films from, *23*:1067
Platinum [*7440-06-4*], *24*:856
    emissivities of, *23*:827
    as refractory coating, *21*:91
    role in temperature measurement,
        23:816
    in shape-memory alloys, *21*:965, 970
    vapor pressure of, *23*:1046
Platinum–rhenium
    catalysts, *21*:342
Platinum–rhodium
    thermocouples, *23*:823
PlazJet, *21*:90
Plictran, *24*:139
Plumbostannite, *24*:123
Plummer's disease, *24*:91
Plums, *23*:96
Plurafac
    ethoxylated alcohol surfactant, *23*:508
Pluronic polyol, *23*:522 Plusride, *21*:33
Ply felts
    as roofing material, *21*:439
PMA. See *Premarket Approval
    Application.*
PMDI. See *Polymeric isocyanate.*
PMN. See *Premanufacture Notification.*
pMOSFETs, *21*:742
Pneumatic atomizer, *22*:672
Pneumatic tire, *24*:161
Pneumococcal pneumonia
    vaccines against, *24*:733
p–n Junction, *21*:732
Pocket calculators
    silver cmpds for batteries, *22*:190

POE. See *Polyolester.*
Poiseuille equation, *21*:310
Polarimetry
    of sugar solutions, *23*:12
Polarizability
    role in optical spectroscopy, *22*:648
    supercritical fluids, *23*:455
Polarization
    measurements by scattering techniques,
        22:648
    of sugar solutions, *23*:13
Polarography
    DMSO as solvent for, *23*:228
Polectron, *24*:1094
Poliglecaprone
    as sutures, *23*:542
Poliglecaprone-25, *23*:548
Polio
    vaccines against, *24*:727
Polishes
    silica gels in, *21*:1000
    triarylmethane dyes for, *24*:565
Pollucite [*1308-53-8*]
    rubidium in, *21*:593
Pollutants
    nir analysis in water, *22*:641
Polyacrylamide grout, *22*:456
Polyacrylamides, *21*:1019
    for hydroseeding, *22*:461
    in soil conditioning, *22*:458
Polyacrylates
    thiols in, *24*:29
Poly(acrylic acid)
    membrane for ultrafiltration, *24*:606
    sol–gel silica hybrids of, *22*:526
    water additive in steam prdn, *22*:740
Polyacrylonitrile
    microfibrils of, *23*:473
Polyacrylonitrile fibers
    dyes for, *24*:552
Polyamides
    membrane for ultrafiltration, *24*:604
    in reverse osmosis membranes, *21*:305
    water additive in steam prdn, *22*:740
Polyamides, aromatic
    as reverse osmosis membranes, *21*:304
Polyaromatic hydrocarbons
    removal from soil, *23*:466
Poly(arylene ether) ketone/silica
    use in sol–gel technology, *22*:526
Polybasite [*53810-31-4*], *22*:170
Polybutadiene, *21*:485

Polybutadiene–acrylonitrile
  thiols in, *24*:28
Polybutadiene-based urethanes
  as sealants, *21*:657
Polybutene olefin sulfonate, *23*:166
Polybutester
  as sutures, *23*:542
  sutures from, *23*:551
Poly(*n*-butylacrylate)
  fractionation using SCFs, *23*:469
Polybutylene adipate (polybutilate),
  23:550
Poly(butylene terephthalate)
  polyesterification of, *24*:326
  as sutures, *23*:542
Poly(*p-tert*-butylstyrene) [*26009-55-2*]
  glass-transition temperature, *22*:1024
Poly(*n*-butyl vinyl ether) [*25232-87-5*],
  24:1059
Poly(*t*-butyl vinyl ether) [*25655-00-9*],
  24:1059
Polycaprolactam, *24*:163
Polycaprolate
  suture coating, *23*:547
Polycarbonates
  resin formation, *24*:326
  silane coupling agent for, *22*:152
  sol–gel–silica hybrids of, *22*:526
Polycarbosilane
  SCF solution of, *23*:472
Polycarboxylic acids
  in textile finishing, *23*:904
Polycat, *24*:700
Polychlorinated biphenyls (PCBs), *21*:166;
  24:492
  removal from soil using SCF, *23*:466
Poly(*p*-chlorostyrene) [*24991-47-7*]
  glass-transition temperature, *22*:1024
Polychlorotrifluoroethylene
  critical surface tension of, *22*:148
Polyclar, *24*:1094
Polyclonal antibodies, *24*:508
Polycyclic aromatic hydrocarbons
  trace analysis of, *24*:495
Polydextrose [*68424-04-4*], *23*:558
Poly(1,1-dihydroperfluorooctyl
    methacrylate) [*29014-57-1*]
  release substrate, *21*:212
Poly(*N,N*-dimethylamide)
  sol–gel silica hybrids of, *22*:526
Poly(*trans*-2,5 dimethyl)piperazinthio-
    furazanamide
  for reverse osmosis membranes, *21*:315

Polydimethylsiloxane
  in supercritical carbon dioxide, *23*:460
  suture coating, *23*:550
Polydimethylsiloxane [*9016-00-6*]
  as release agent, *21*:210
  sol–gel hybrids of, *22*:525
Polydimethylsiloxane film
  release substrate, *21*:212
Polydimethylsiloxane fluid
  release substrate, *21*:212
Poly(2,4-dimethylstyrene) [*25990-16-3*]
  glass-transition temperature, *22*:1024
Poly(2,5-dimethylstyrene) [*34031-72-6*]
  glass-transition temperature, *22*:1024
Polydioxanone, *23*:548
  as sutures, *23*:542
Polyester, *23*:543
  silane coupling agent for, *22*:152
  tire cords, *24*:162, 163
  titanate catalysis, *24*:325
Polyester polyols
  polyurethane products, *24*:696
Polyester urethanes
  as sealants, *21*:657
Polyether polyols
  polyurethane products, *24*:696
Polyether urethanes
  as sealants, *21*:657
Polyethylbenzenes (PEB), *22*:959
Polyethylene [*9002-88-4*]
  critical surface tension of, *22*:148
  release substrate, *21*:212
  silane coupling agent for, *22*:152
  supercritical fluids for prdn, *23*:465
  as sutures, *23*:542
Poly(ethylene glycol)
  in soil conditioning, *22*:458
  for sulfur recovery, *23*:443
Polyethylene glycol dimethyl ether, *23*:440
Polyethyleneimine, *21*:1019
Polyethylene ketone
  in tire cord, *24*:164
Poly(ethylene oxide)/silica
  sol–gel hybrids of, *22*:526
Poly(ethylene terephthalate)
  critical surface tension of, *22*:148
  polyesterification of, *24*:326
Polyethylene terephthalate [*25038-59-9*],
  23:550
Polyethylene wax [*8002-88-4*]
  as release agent, *21*:210
Poly(2-ethylhexyl vinyl ether) [*29160-05-
  2*], *24*:1059

Poly(ethyl methacrylate)
 SCF solution of, 23:472
Poly(ethyloxazoline)/silica
 use in sol–gel technology, 22:526
Poly(ethyl vinyl ether) [25104-37-4],
  24:1059
Poly(fluorethers)
 as release agent, 21:210
Polyfuran, 21:318
Polygallitol, 23:96
Polyglactin
 as sutures, 23:542
Polyglactin-910 [26780-50-7]
 sutures of, 23:547
Polyglycolic acid [26124-68-5]
 as sutures, 23:542
 sutures of, 23:546
Polyglycols
 water additive of steam prdn, 22:740
Polyglyconate, 23:548
 as sutures, 23:542
Polyharnstoff dispersion (PHD) polyols,
  24:709
Poly(n-hexyl vinyl ether) [25232-88-6],
  24:1059
Polyhydroxyamides, 23:521
Polyimide
 in sol–gel polymer hybrids, 22:526
Poly(isobutyl vinyl ether) [9003-44-5],
  24:1059
Poly(isobutyl vinyl ether-co-vinyl chloride)
 [25154-85-2], 24:1064
Polyisocyanates, 24:285
Polyisocyanurates
 titanate catalysts for, 24:289
cis-1,4-Polyisoprene
 natural rubber, 21:572
Poly(isopropyl vinyl ether) [25585-49-3],
  24:1059
Polylactic acid [26023-30-3], 23:546
Poly(L-lactic acid), 23:472
Polymer chemistry, 24:164
Polymer composites, 21:194
Polymer emulsions
 recycling, 24:957
Polymer grouts, 22:454
Polymeric isocyanate (PMDI), 24:696
Polymeric materials
 as roofing material, 21:446
Polymeric plasticizers, 24:1034
Polymeric sucrose derivatives, 23:75
Polymerization
 DMSO as solvent for, 23:227

SCFs, 23:465
 sulfolane as solvent, 23:139
 of vinyl acetate, 24:944
Polymer processing
 effect of supercritical fluids, 23:462
Polymers, 23:472
 analysis by infrared spectroscopy,
  22:641
 rheological measurements, 21:427
 selenium-containing, 21:702
 from sucrose, 23:7
 sulfolane as solvent for, 23:139
 of sulfur trioxide, 23:154
 in supercritical fluids, 23:460
 thin films of, 23:1073
 use of hexitols in, 23:107
Polymetallocarbosilanes, 24:303
Polymetatelluric acid, 23:798
Polymethacrylates, 24:29
 thiols in, 24:29
Polymethylene polyphenyl isocyanate
 [9016-87-9], 24:706
Polymethylhydrosiloxane [9004-73-3],
  22:50, 57
 as waterproofing agent, 22:47
Poly(methyl methacrylate)
 effect of supercritical carbon dioxides,
  23:461
 rubber grafting, 21:575
 SCF solution of, 23:472
 sol–gel silica hybrids of , 22:526
 supercritical carbon dioxide as foaming
  agent, 23:460
Poly[(methyl methacrylate)-co-styrene]
 effect of supercritical carbon dioxides,
  23:461
Polymethyl(3,3,4,4,5,5,6,6,6-nonafluoro-
  hexyl)siloxane
 release substrate, 21:212
Polymethyl(nonafluorohexyl)siloxane
 [115287-18-8]
 as release agent, 21:210
Poly(m-methylstyrene) [25037-62-1]
 glass-transition temperature, 22:1024
Poly(o-methylstyrene) [25087-21-2]
 glass-transition temperature, 22:1024
Poly(p-methylstyrene) [24936-41-2]
 glass-transition temperature, 22:1024
Poly(α-methylstyrene) [25014-31-7]
 glass-transition temperature, 22:1024
Poly(methyl vinyl ether) [34464-52-6],
  24:1059

Poly[(methyl vinyl ether)-*co*-maleic anhydride] [*9011-16-9*], *24*:1064

Poly(octadecyl vinyl ether-*co*-maleic anhydride) [*28214-64-4*], *24*:1064

Poly(*n*-octyl vinyl ether) [*25232-89-7*], *24*:1059

Polyolefins, *21*:864
   as release agents, *21*:210

Polyolester (POE)
   as compressor oil, *21*:137

Polyols, *23*:93; *24*:296

Polyoxalkylenes
   as release agent, *21*:210

Polyoxazolidones
   titanate catalysts for, *24*:289

Poly[oxy(dimethylsilylene)] [*9016-00-6*], *22*:110

Poly(oxyethylene(20)) sorbitan monooleate, *23*:110

Poly[oxy(methylphenylsilylene)] [*9005-12-3*], *22*:110

Poly[oxy(methyl-3,3,3-trifluoropropyl)-silylene] [*25791-89-3*], *22*:110

Polyozoline/silica
   sol–gel hybrids of, *22*:526

Poly(*n*-pentyl vinyl ether), *24*:1059

Polypeptide, *22*:150

Polyperfluoroalkylene ethers
   titanium complex of, *24*:300

Polyphenolic tannins
   as bird repellents, *21*:254

Polyphenol oxidase
   in teas, *23*:751

Polyphenols
   in fermented teas, *23*:751

Poly(*p*-phenylene benzobisoxazole)
   in tire cord, *24*:164

Poly(phenylene sulfide), *22*:414

Polyplasdone, *24*:1094

Polypropylene [*9003-07-0*], *21*:864
   critical surface tension of, *22*:148
   as release agent, *21*:210, 212
   silane coupling agent for, *22*:152
   as sutures, *23*:542
   sutures of, *23*:551
   for tea packaging, *23*:760

Poly(propylene terephthalate)
   polyesterification of, *24*:326

Polysaccharides
   dextrose in prdn of, *23*:593
   in soil conditioning, *22*:458

Polysilanes, *22*:33

Polysilicic acid, *21*:1009

Polysilicon, *23*:1066

Polysiloxanes
   cured by titanates, *24*:323
   role in self-assembled monolayer formation, *23*:1088

Polysorb, *23*:543

Polysorbate 80, *23*:111

Polystyrene (PS) [*9003-53-6*], *22*:956, 984, 1015
   critical surface tension of, *22*:148
   effect of supercritical carbon dioxides, *23*:461
   glass-transition temperature, *22*:1024
   silane coupling agent for, *22*:152
   sol–gel silica hybrids of, *22*:525
   thiols in, *24*:29

Polysulfides
   formation of, *23*:429

Polysulfide sealants, *21*:658

Polysulfones
   membrane for ultrafiltration, *24*:604
   for reverse osmosis membranes, *21*:315

Polytellurides, *23*:786

Polyterephthalate
   silane coupling agent for, *22*:152

Polytergent
   ethoxylated alcohol surfactant, *23*:508
   ethoxylated alkylphenol surfactant, *23*:511

Polyterpene, *23*:833

Polytetrafluoroethylene [*9002-84-0*], *23*:550
   critical surface tension of, *22*:148
   as release agent, *21*:210
   release substrate, *21*:212
   as sutures, *23*:542

Polytetramethylene ether glycol [*25190-06-1*], *23*:551

Poly(tetramethylene oxide)
   sol–gel ceramic hybrids of, *22*:526

Polythionic acids, *24*:53

Polythiophenes, *24*:48

Polytitanates, *24*:251

Polytitanocarbosilanes, *24*:303

Polytitanosiloxane, *24*:303

Polytitanoxanes, *24*:282

Poly(urea–formaldehyde)
   in soil conditioning, *22*:458

Polyureas, *23*:271
   in reverse osmosis membranes, *21*:305

Polyurethane (PU). See *Urethane polymers.*

Polyurethane elastomers, 24:724
Polyurethane foams
  organotin catalysts for, 24:145
Polyurethane-modified isocyanurate
  (PUIR), 24:696, 697
Polyurethane resins, 23:112
Polyurethanes, 24:695
  in soil conditioning, 22:458
Polyurethane surface coatings, 24:724
Poly(vinyl acetal). See Vinyl acetal
  polymers.
Poly(vinyl acetal) resin, 24:924
Poly(vinyl acetate) [9003-20-7], 24:943
  hydrolysis of, 24:980
Poly(vinyl alcohol) (PVA) [9002-89-5],
  24:291, 980
  acetalization of, 24:924
  cross-linking with titanates, 24:331
  for hydroseeding, 22:461
  as release agent, 21:210
  in soil conditioning, 22:458
  sol–gel silica hybrids of, 22:525
  in tire cord, 24:164
Poly(vinyl alcohol)/silica
  use in sol–gel technology, 22:526
Poly(vinyl benzal), 24:924
Poly(vinyl butyral) (PVB) [63148-65-2],
  24:924
  from PVA, 24:980
Poly(vinyl carbonates)
  from poly(vinyl alcohol), 24:990
Poly(vinyl chloride) (PVC) [9002-86-2],
  21:864; 24:851
  critical surface tension of, 22:148
  recycling of, 24:1044
  as roofing material, 21:448
  silane coupling agent for, 22:152
  thiols in, 24:29
  use of thioglycolic acid in, 24:13
Poly(vinyl chloride) resins
  sorbitol as stabilizer for, 23:112
Poly(vinyl ethers), 24:1053
Poly(vinyl fluoride) [24981-14-4], 24:1099
  release substrate, 21:212
Poly(vinyl formal) (PVF) [9003-33-2],
  24:924, 936
Poly(vinylidene chloride), 24:882
Poly(vinylidene fluoride) [24937-79-9],
  24:923
  membrane for ultrafiltration, 24:604
  release substrate, 21:212
Poly(vinyl ketals), 24:924

Poly(vinyl nitrate), 24:989
Poly(vinyl phosphate), 24:990
Polyvinylpyrrolidinone, 24:1070
  sol–gel silica hybrids of , 22:526
Poly(N-vinyl-2-pyrrolidinone) [9003-39-8],
  24:1070
Pontianak [9000-14-0], 21:297
Poole-Frenkel mechanism, 23:673
Popcorn polymerization
  of N-vinyl-pyrrolidinone, 24:1077
Porcelain
  selenium as colorant for, 21:712
  sodium nitrate for prdn of, 22:393
Poromeric materials
  polyurethanes, 24:724
Portland cement
  sulfuric acid as by-product, 23:395
Positron emission tomography
  sodium iodide detectors for, 22:382
Potassium
  complexes with sugar alcohols, 23:103
  diffusion coefficient in vitreous silica,
  21:1047
  in tea, 23:751
  vitreous silica impurity, 21:1036
Potassium acid phthalate
  use in x-ray spectroscopy, 22:654
Potassium channels
  role in sweetness, 23:578
Potassium chloride, 22:366
Potassium citrate
  use in density gradient separation,
  21:853
Potassium dititanate [12056-46-1], 24:251
Potassium fluorotitanate [23969-67-7],
  24:256
Potassium heptafluorotantalate [16924-
  00-8], 23:663, 675
Potassium hexachlorotitanate [16918-46-
  0], 24:261
Potassium hexacyanoferrate(III), 24:500
Potassium hydroxide [1310-58-3], 24:854
Potassium p-hydroxybenzoate [16782-08-
  4], 21:621
Potassium iodate
  in table salt, 22:373
Potassium iodide
  in table salt, 22:373
Potassium metatitanate [12030-97-6],
  24:251
Potassium metavanadate [13769-43-2],
  24:800

Potassium nitrate
  from sodium nitrate, 22:388
Potassium permanganate [7722-64-7],
  24:855
Potassium polytitanate [12056-49-4],
  24:251
Potassium salicylate [578-36-9], 21:612
Potassium selenocyanate [3425-46-5],
  21:698
Potassium silicate [1312-76-1], 22:1
Potassium–sodium alloys, 22:347
Potassium sodium nitrate
  fertilizer, 22:389
Potassium sorbate [590-00-1], 22:572
E,E-Potassium sorbate [24634-61-5],
  22:572
Potassium sulfamate [13823-50-2], 23:125
Potassium tellurate [7790-58-1], 23:799
Potassium tetrathiotungstate [14293-75-
  5], 24:597
Potassium thiocyanate [333-20-0], 23:322
Potassium thioglycolate [34452-51-2], 24:6
Potassium thiosulfate [10294-66-3], 24:55
Potassium titanium oxide [12056-51-8],
  24:251
Potato
  dextrose from, 23:585
  processing, 21:327
  starch from, 22:699
Potato chips
  oil removal from, 23:467
Potato starch, 22:714
  centrifugal separation of, 21:867
Potency
  of vaccines, 24:732
Potentially responsible party (PRP),
  21:166
POTW. See Publicly owned treatment
  work.
Poultry
  use of sorbates, 22:583
Povidone-iodine, 24:1088
Powder clustering, 22:235
Powdered sugar, 23:41
Powder metallurgy
  hypereutectic Al–Si alloys, 21:1109
  of rhenium and rhenium alloys, 21:340
  use for tool materials, 24:401
Powders
  particle size enlargement, 22:222
  rheological measurements, 21:427
  talc in, 23:614

Power
  remote loads, 22:475
  sulfuric acid by-product, 23:379
Power cycles, 22:721
Power generation
  regulatory agencies, 21:179
  steam for, 22:746
  thermoelectricity for, 23:1029
Power plants
  integrated gasification combined cycle,
    23:435
  shape-memory alloys for, 21:969
Power semiconductors
  silicon-based semiconductors, 21:744
Pralidoxime chloride, 24:830
Prandtl number, 22:199
Praziquantel [55268-74-1], 24:832
Precalciferol [50524-96-4], 22:857
Precipitants
  heavy metal, 22:414
Precision casting
  ethoxysilanes for, 22:78
Prednisolone [50-24-8], 22:884
Prednisolone, synthetic [52438-85-4],
  24:832
Prednisone [53-03-2], 22:884, 905; 24:837
Pregelatinized starches, 22:712
Pregnane [24909-91-9], 22:854
Pregnenolone [145-13-1], 22:871
$17\eta H$-Pregn-5-en-20-yne-3η-dio [3604-
  60-2], 22:886
Preheaters
  sulfamic acid cleaner for, 23:129
Premanufacture notification (PMN),
  21:166
Premarket Approval Application (PMA),
  21:175
Prenyl chloride [503-60-6], 23:836
Prepregs, 21:197
Preservatives
  p-hydroxybenzoic acid in, 21:620
  sodium chloride as, 22:372
Pressure atomizer, 22:672
Pressure bag molding, 21:197
Pressure compaction, 22:229
Pressure measurements
  sensors for, 21:820
Pressure-sensitive adhesives (PSA)
  silicone products, 22:116
Pressure-vacuum valve
  role in tank design, 23:630
Pressure vessels, 23:631

Prevention of Significant Deterioration (PSD), *21*:166

Primers for metals
  PVB in, *24*:935

Primrose
  sugar alcohols in, *23*:96

Printers' rollers
  sorbitol in, *23*:112

Printing inks, *23*:208
  alkyl titanates in, *24*:328
  PVC emulsions for, *24*:1030
  rheological measurements, *21*:427
  solvents for, *22*:570
  from triarylmethane dyes, *24*:557

Prisms, *21*:1083
  of vitreous silica, *21*:1065

Proanthocyanidins
  as bird repellents, *21*:254

Procaine hydrochloride [*51-05-8*], *24*:835

Process control
  size measurement of particles for, *22*:264

Process cooling, *21*:128

Processed cheeses
  sorbates in, *22*:582

Process heating
  from steam, *22*:755

Processing agents, *21*:478

Process iron, *23*:391

Process monitoring
  role of optical spectroscopy, *22*:638

Producer gas
  steam in prdn of, *22*:758

Product safety regulation, *21*:162

Profenophos
  thiols in, *24*:30

Progesterone [*57-83-0*], *22*:860; *24*:833

Progestins, *22*:852

Prolactin [*12585-34-1*], *24*:833

Prolene, *23*:543

Proliferous polymerization
  of *N*-vinyl-pyrrolidinone, *24*:1077

Proline
  in silk, *22*:156
  in soybeans and other oilseeds, *22*:596

Promethazine [*60-87-7*], *22*:941

Prometon [*1610-18-0*], *22*:425

Prometryne, *24*:510

Propane [*74-98-6*]
  purification of, *23*:268
  as SCF, *23*:468
  supercritical fluid, *23*:454

Propane cannons
  as bird repellents, *21*:254

Propanesulfonic acid [*28553-80-2*], *23*:194

1-Propanethiol [*107-03-9*], *24*:21

2-Propanethiol [*75-33-2*], *24*:21

Propanil [*709-98-8*], *22*:422

1-Propanol [*71-23-8*]
  as solvent, *22*:536

2-Propanol [*67-63-0*]
  as solvent, *22*:536

2-Propanol–water system, *21*:949

Propargite [*2312-35-8*]
  from thionyl chloride, *23*:295

Propazine, *24*:510

Propionibacterium freundenreichii
  inhibited by sorbates, *22*:580

Propionibacterium zeae
  inhibited by sorbates, *22*:580

Proposition 65, *22*:532

n-Propyl acetate [*109-60-4*]
  as solvent, *22*:542

n-Propyl amine [*107-10-8*]
  as solvent, *22*:540

n-Propyl chlorosulfate [*819-52-3*], *23*:411

n-Propyl chlorosulfite [*22598-38-5*], *23*:411

Propylene [*115-07-1*], *24*:866

Propylene carbonate [*108-32-7*]
  as solvent, *22*:544

Propylene dichloride [*78-87-5*]
  as solvent, *22*:538

Propylene glycol [*57-55-6*], *23*:109
  salt additive, *22*:368
  as solvent, *22*:546

Propylene glycol monobutyl ether [*5131-66-8*]
  as solvent, *22*:544

Propylene glycol monoethyl ether [*1569-02-4*]
  as solvent, *22*:544

Propylene glycol monomethyl ether [*1589-49-7*]
  as solvent, *22*:544

Propylene glycol monomethyl ether acetate [*108-65-6*]
  as solvent, *22*:544

Propylene glycol monophenyl ether [*770-35-4*]
  as solvent, *22*:546

Propylene glycol monotertiary butyl ether [*57018-52-7*]
  as solvent, *22*:544

Propylene oxide [75-56-9]
    silver cmpd catalysts for, 22:190
    as solvent, 22:548
1,3-Propylene sulfate [1073-05-8], 23:410
1,2-Propylene sulfite [1469-73-4], 23:410
n-Propyl propionate [106-36-5]
    as solvent, 22:542
6-Propyl-2-thiouracil [51-52-5], 24:100
Propyltriethoxysilane [141-57-1], 22:72
Propyltrimethoxysilane [1067-25-0], 22:72
    critical surface tension of, 22:148
Prostaglandins
    silylation in synthesis of, 22:146
    as veterinary drugs, 24:833
Prosthetics
    shape-memory alloys for, 21:973
    tantalum in, 23:671
Protective agents
    for rubber, 21:574
Protective colloids
    PVA resins as, 24:980
Protein dispersibility index, 22:613
Protein meals
    of soybeans and other oilseeds, 22:612
Proteins
    centrifugal separation, 21:852
    crystal growth in space, 22:621
    effect of SCFs on, 23:467
    fluorescent probe for, 23:211
    nir spectroscopy of, 22:641
    in silk, 22:156
    in soybean and other oilseeds, 22:594
    ultrafiltration, 24:622
Proteus vulgaris
    inhibited by sorbates, 22:580
Prothiophos
    thiols in, 24:30
Protocatechaic acid [99-50-3], 24:813
Protriptyline [438-60-8], 22:942
Proustite [15152-58-4], 22:170, 179
Provitamin A, 23:875
Prozac, 22:945
PRP. See Potentially responsible party.
Prunes
    sorbates in, 22:583
Pruritis
    selenium cmpds for, 21:713
Prussian Blue
    salt additive, 22:368
PS. See Polystyrene.
PSA. See Pressure-sensitive adhesives.
PSD. See Prevention of Significant
    Deterioration.

Pseudoelasticity
    of shape-memory alloys, 21:963
Pseudoionone [141-10-6], 23:864
Pseudomonad plasmid
    salicylic acid prprn, 21:609
Pseudomonas fluorescens
    inhibited by sorbates, 22:580
Pseudomonas fragi
    inhibited by sorbates, 22:580
Pseudomonas sp.
    inhibited by sorbates, 22:580
Pseudorutile [1310-39-0], 24:241
PS foams, 22:1061
PS ionomers, 22:1021
Psychoanaleptics, 22:935
Psychostimulants, 22:935
PTC resistors
    barium titanate for, 24:253
PT diagram
    in thermodynamics, 23:989
PTE. See Passenger tire equivalents.
PU. See Polyurethane.
Publicly owned treatment work (POTW),
    21:166
Public Utility Regulatory Policy Act
    (PURPA), 21:181
PUIR foams, 24:697
Pulleys
    magnetic for separations, 21:878
Pullulanase, 23:589, 599
Pullularia pullulans
    inhibited by sorbates, 22:579
Pulp
    reverse osmosis for wastewater, 21:322
    steam system for, 22:754
    use of sulfur dioxide, 23:310
    use of sulfuric acid, 23:397
    use of ultrafiltration, 24:622
Pulp and paper mills
    scrap tire fuel, 21:25
Pulping
    in paper recycling, 21:12
    sodium sulfite used in, 23:314
Pulp-mill wastes
    thickening of, 21:680
Pultrusion, 21:201
Pummerer reactions, 23:221
Pumping speed
    in vacuum systems, 24:769
Pumpkin
    mannitol in, 23:97
Pumps, high pressure oxygen
    silver parts for, 22:168

Pure ethyl silicate, *22*:76

Pure & Natural, *22*:323

Pure silicon
   trichlorosilane in mfg of, *22*:36

PURPA. See *Public Utility Regulatory Policy Act.*

Purple glass, *21*:712

PV. See *Photovoltaic cells.*

PVA. See *Poly(vinyl alcohol).*

PVA resins, *24*:980

PVB. See *Poly(vinyl butyral).*

PVC. See *Poly(vinyl chloride).*

PVC stabilizers
   thiols for, *24*:30

PV diagram
   in thermodynamics, *23*:989

PVF. See *Poly(vinyl formal).*

PVF resins, *24*:936

PVP hydrogels, *24*:1078

Pyrantel [*15686-83-6*], *24*:48

Pyrantel pamoate [*22204-24-6*], *24*:832

Pyrargyrite [*15123-77-0*], *22*:170, 179

Pyrene
   analysis using laser-induced
      fluorescence, *22*:654

Pyridate
   thiols in, *24*:30

Pyridine [*110-86-1*]
   regulatory level, *21*:160
   silver complexes of, *22*:186

(4-Pyridyl)thioacetic acid [*10351-19-8*],
   24:16

Pyrite [*1309-36-0*], *23*:303
   selenium in, *21*:689
   sulfide ore, *23*:244
   tellurium in, *23*:782

Pyrocatechol, *21*:1088

Pyroceram, *24*:323

Pyrochlore
   magnetic intensity, *21*:876

Pyrodextrins, *22*:710

Pyrolusite
   magnetic intensity, *21*:876

Pyrolysis
   of scrap tires, *21*:28

Pyrometer, optical
   role in temperature measurement,
      23:824

Pyrometers
   detectors for beam spectroscopy, *22*:657

Pyroselenic acid [*14998-61-9*], *21*:701

Pyrosulfuryl chloride [*7791-27-7*], *23*:286

Pyrosultone, *23*:160

Pyrotechnics
   strontium compounds in, *22*:954

Pyrrhotite [*1310-50-5*]
   magnetic intensity, *21*:876
   sulfide ore, *23*:244
   tellurium in, *23*:782

Pyrroelectric coefficient
   of thin films, *23*:1086

Q

QED. See *Quantum-effect devices.*

Q Fiber, *21*:124

Quadrol [*102-60-3*], *24*:296

Quantum cellular automata, *21*:786

Quantum-effect devices (QEDs), *21*:785

Quantum well infrared photodetector
   (QWIP), *21*:797

Quartz [*14808-60-7*], *21*:981, 1087
   in silica refractories, *21*:83
   substrate for self-assembled monolayer,
      23:1089
   thin films of, *23*:1055
   use in optical spectroscopy, *22*:635, 644

α-Quartz, *21*:1076

Quartz crystal microbalances
   sensors based on, *21*:823

Quartz crystal oscillator
   use in thin films, *23*:1048

Quartz glass
   as sampling probe, *21*:634

Quasi-elastic light scattering
   for particle size measurement, *22*:270

Quaternary ammonium chlorides
   silylating agent on, *22*:150

Quaternary ammonium salts, *22*:847;
   23:528

Quaternary ammonium starches, *22*:714

Quebrachitol, *23*:98

Quenching
   of steel, *22*:802

Quercetin [*117-39-5*]
   in tea, *23*:749

Quick-Sep, *24*:493

Quilon, *21*:214

Quinic acid
   in sunflower meal, *22*:598

Quinoline
   methylation of, *23*:223

para-Quinone dioxime [*105-11-3*]
    cross-linking agent, *21*:470
QWIP. See *Quantum well infrared
    photodetector.*

R

RACT. See *Reasonably Available Control
    Technology.*
Radial passenger tires, *24*:176
Radial tires, *21*:493
Radiation-curable vinyl ether coatings,
    24:1066
Radiation-induced polymerization
    of cyclic siloxanes, *22*:94
Radiators
    brazing filler metals for, *22*:490
Radioactive dating
    use of rubidium compounds, *21*:599
Radioactive processing
    wastewater treatment for, *21*:320
Radioactive tracers
    rubidium as, *21*:598
    in trace analysis, *24*:493
Radioactive waste
    from power generation, *21*:191
Radio astronomy
    role of optical spectroscopy, *22*:636
Radioisotopes, *23*:1035
    for sterilization, *22*:845
Radiology
    use of tantalum in, *23*:671
Radiometry, *22*:641
Radios
    synthetic quartz crystals for, *21*:1082
Raffinose
    in beet sugar, *23*:55
    in soybeans and other oilseeds, *22*:598
Rafoxanide [*22662-39-1*], *24*:832
Railroads, *24*:526
Rainwater, *23*:217
Raloxifene [*84449-90-1*], *22*:904
Raman effect
    role in optical spectroscopy, *22*:648
Raman-induced Kerr-effect spectroscopy
    (rikes), *22*:650
Raman scattering, *21*:1035; *22*:628
Raman spectroscopy, *22*:516
Rancimat system, *23*:763
Raney nickel
    catalysis for sugar alcohols, *23*:105

Rankine cycle, *22*:749
Rank Pulse Shearometer, *21*:424
Raoult's law, *22*:196, 202; *23*:1019, 1045
Rape oil
    specific gravity, *23*:625
Rapeseed
    extraction using SCFs, *23*:467
Rapid expansion from supercritical
    solutions (RESS), *23*:471
Rare-earth alloys
    role in steelmaking, *22*:780
Rare earths, *21*:1118
    silicides, *21*:1118
Raspberries
    sulfoxides in, *23*:217
Ray diagram, *22*:728
Rayleigh scattering, *22*:628
Rayon
    sodium sulfide in prdn of, *22*:417
    in tire cord, *24*:162
    tire cords, *24*:162
    use of sulfur for, *23*:251
Rayon cake, *22*:403
RCF. See *Refractory ceramic fibers.*
RDF. See *Refuse-derived fuel.*
R&D management, *21*:265
Reaction injection molding (RIM), *21*:198
Reaction mechanisms
    for MOCVD, *21*:773
Reactive-ion etching (RIE), *21*:800
Reactive-ion plating
    of refractory coatings, *21*:96
Reactive sputtering
    of refractory coatings, *21*:96
Reactors
    tantalum liners, *23*:671
Reasonably Available Control Technology
    (RACT), *22*:531
Rebaudioside A [*58543-16-1*]
    as sweetener, *23*:567
Rebound flexible foam, *24*:719
Reclaiming
    of oil, *21*:6
    rubber, *21*:22, 492
Rectifier, *21*:732
    selenium, *21*:708
Rectifying junction, *21*:735
Rectisol, *23*:440
Recycle Selectox process, *23*:445
Recycling
    of asphalt materials, *23*:261
    ferrous metals, *20*:**1092**

glass, *20*:**1127**
    magnetic separators for, *21*:898
    nonferrous metals, *20*:**1092**
    oil, *21*:**1**
    paper, *21*:**10**
    plastics, *21*:22
    rubber, *21*:**22**
    water, *21*:46
Red lead [*1314-41-6*]
    as rubber chemical, *21*:473
Red mud
    thickening of, *21*:681
Red seaweed, *23*:98
Red tide algae, *24*:505
Reduced viscosity, *21*:356
Reducing sugars
    determination of, *23*:15
Reductive bleaching
    in paper recycling, *21*:18
Red wine
    use of sorbates, *22*:582
Reed reaction, *23*:161, 302, 494
Refinery molasses, *23*:42
Refining
    in paper recycling, *21*:18
Refining, petroleum. See *Petroleum refining.*
Reflective tapes
    as bird repellents, *21*:254
Reflectivities
    of white pigments, *24*:240
Refraction, *21*:46
Refractive index
    in sugar analysis, *23*:14
Refractories, *21*:**46**
    effect of silicate glass melts, *22*:12
    silica for, *21*:999
    silicon esters in prprn of, *22*:79
    talc application, *23*:613
    thorium as, *24*:71
    titanium silicides in prprn of, *24*:263
Refractoriness, *21*:77
Refractory brick
    use in ladle metallurgy, *22*:780
Refractory ceramic fibers (RCF), *21*:125
Refractory coatings, *21*:**87**
Refractory fibers, *21*:**117**
Refractory metals
    by sodium reduction, *22*:344
Refrasil, *21*:123
Refrigeration, *21*:**128**
    thermoelectricity for, *23*:1029

Refrigeration systems
    role of thermoelectric energy, *23*:1038
Refrigerator/freezers, *21*:128
Refuse
    fuel for steam prdn, *22*:746
Refuse-derived fuel (RDF), *21*:229
Regenerated cellosic fibers, *21*:149
Reg Neg process, *22*:533
Regulatory agencies, *21*:**149**
    chemical process industry, *21*:**154**
    pharmaceuticals and cosmetics, *21*:**168**
    power generation, *21*:**179**
Reich test, *23*:308
Reinforced plastics, *21*:**194**
    silylating agents for, *22*:151
Reinforced RIM, *21*:198
Reinforced styrene polymers, *22*:1028
Reinforcement media
    textile use, *23*:882
Relative viscosity, *21*:356
Release agents, *21*:**207**; *23*:208
Relief pads
    from ground rubber, *21*:35
Relishes
    sorbates in, *22*:583
Remote sensing
    by infrared spectroscopy, *22*:641
Renewability
    of surfactants, *23*:532
Renewable energy resources, *21*:**218**
Renex
    polyoxyethylene surfactant, *23*:514
Renierite
    magnetic intensity, *21*:876
Repel, *21*:258
Repellents, *21*:**236**
Reportable quantity (RQ), *21*:166
Reprocessing
    of oil, *21*:6
Reprographic inks
    stilbene dyes in, *22*:929
Reprography, *21*:263
Re-refining
    of oil, *21*:7
Research management, *21*:263
Research/technology management, *21*:**264**
Resenes, *21*:297
Reserpine [*50-55-5*], *22*:939
Residual gas analyzers
    for vacuum systems, *24*:773
Residual properties
    in thermodynamics, *23*:1005

Residual property relation
   of thermodynamics, 23:1016
Residue analysis, 24:491
Residue curve maps
   in separation process synthesis, 21:929
Resin, 23:719
   fractionation using SCFs, 23:469
   regeneration using SCFs, 23:468
   sampling standards for, 21:627
   silylating agents for interfaces, 22:151
   titanate catalysis for, 24:326
   use in refractory brick, 21:69
   water-soluble, 21:303
Resin acids
   in tall oil, 23:616
Resin component, 24:168
Resinols, 21:297
Resins, natural, 21:**291**
Resin-transfer molding (RTM), 21:197
Resin treatments
   in textile finishing, 23:891
Resistance
   role in temperature measurements,
      23:818
Resistant starch, 22:702
Resistazone counter, 22:275
Resistors
   refractory coatings for, 21:113
Resolution
   role in optical spectroscopy, 22:632
Resonant ionization mass spectrometry,
   22:658
Resonantly enhanced multiphoton
   ionization, 22:657
Resonant tunneling diode (RTD), 21:785
Resonators
   high frequency, 23:677
Resorcinol [108-46-3], 21:303, 479; 24:79
   sulfonic acid in prdn of, 23:206
Resorcinol–formaldehyde– latex, 24:163
Resource Conservation and Recovery Act,
   21:166
Respiratory syncytial virus
   vaccine for, 24:736
Responsible Care, 21:154
RESS. See Rapid expansion from
   supercritical solutions.
Restriction endonucleases, 24:516
Retarders
   rubber chemicals, 21:460
   sodium nitrite use in, 22:400
Retreading, 21:43

Retroviruses, 21:303
Reversed-phase liquid chromatography
   for trace and residue analysis, 24:501
Reverse jet scrubber, 23:392
Reverse micelles
   effect of supercritical fluids, 23:462
Reverse osmosis, 21:**303**; 23:467
   sulfonic acid derivatives for, 23:169
   for tea concentration, 23:761
   water treatment for steam prdn, 22:745
Reverse-phase hplc
   of black tea, 23:756
Reverse Rankine cycle
   for refrigeration, 21:129
Reverse transcriptase
   grown in space, 22:622
Rexol
   ethoxylated alkylphenol surfactant,
      23:511
Reynolds number, 21:335, 668
   role in centrifugal separation, 21:838
   role in particle size measurement,
      22:268
   role in reverse osmosis, 21:311
   in sprays, 22:684
L-Rhamnose [3615-41-6], 23:571
Rheniforming process, 24:359
Rhenium [7440-15-5], 21:335
   as refractory coating, 21:91
Rhenium and rhenium compounds, 21:**335**
Rhenium pentachloride [39368-69-9],
      21:343
Rhenium sulfide [1314-68-7], 21:336
Rheological measurements, 21:**347**
Rheology, 21:347
Rheometer, 21:394
Rheopexy, 21:351
Rheosyst, 21:396
Rheotron, 21:394
Rheovibron, 21:419
Rheumatoid arthritis
   treatment of, 24:65
Rhizoctonia solani
   inhibited by sorbates, 22:579
Rhizopus arrhizus
   inhibited by sorbates, 22:579
Rhizopus nigricans
   inhibited by sorbates, 22:579
Rhodacal
   alkylbenzensulfonate surfactant, 23:496
Rhodamine 6G, 22:190
   trace analysis of, 24:492

Rhodapex
    alcohol surfactant, 23:502
    alkylphenol surfactant, 23:503
Rhodapon OLS
    surfactant, 23:501
Rhodasurf
    ethoxylated alcohol surfactant, 23:508
Rhodiarome, 24:822
Rhodium
    as refractory coating, 21:91
Rhodochrosite
    magnetic intensity, 21:876
Rhodonite
    magnetic intensity, 21:876
Rhodotorula flava
    inhibited by sorbates, 22:579
Rhodotorula glutinis
    inhibited by sorbates, 22:579
Rhodotorula rubra
    inhibited by sorbates, 22:579
Rhodotorula spp.
    inhibited by sorbates, 22:579
Rhovanil, 24:815
Ribbed smoked sheet, 21:564
Ribitol [488-81-3], 23:95
Riboflavin, 21:437
Rice
    dextrose from, 23:585
    extraction using SCFs, 23:467
    starch from, 22:699
Rice brand oil
    in soap manufacture, 22:302
Rice hulls, 24:435
    silicon tetrachloride from, 22:35
Richardson and Zaki equation, 21:670
Richonol
    alcohol surfactant, 23:502
Ricinoleic acid, 23:503
Rickardite [12134-39-31]
    tellurium in, 23:783
Rickettsiae
    veterinary drugs against, 24:829
RIE. See Reactive ion etching.
Rifampicin
    for trace and residue analysis, 24:498
Rifle barrels
    refractory coatings for, 21:101
Rigidized thermoplastic sheet
    reinforced plastics, 21:203
Rigid urethane foams, 24:702
Rikes. See Raman-induced Kerr-effect
    spectroscopy.

RIM. See Reaction injection molding.
Rimadyl, 24:832
RIM systems
    for polymethanes, 24:718
Ring-opening polymerization
    of cyclic oligosiloxanes, 22:91
Ringworm
    veterinary drugs against, 24:829
Risk assessment, 21:437
Robenidine [25875-50-7], 24:831
Rocket combustion chambers
    refractory coatings for, 21:106
Rock salt, 22:354
Rockwell hardness
    carbon eutectoid steel, 22:796
    of tool materials, 24:397
Rockwell hardness A
    of cemented carbide tools, 24:406
Rocky Mountain spotted fever, 21:238
Rodenticides, 21:437
Rod mills, 22:287
Roll crushers, 22:285
Roller mills, 22:285
Rolling
    unit operation of tea processing, 23:757
Roll roofing, 21:451
Roofing
    talc for, 23:613
Roofing materials, 21:**437**
Roofing membranes, 21:447
Roofing shingles, 21:450
Roof linings
    refractories for, 21:83
Roofs, 21:437
Room-temperature vulcanizable (RTV)
    silicones, 22:119
Roost-No-More, 21:255
Root control preparations
    in grouts, 22:456
Roots
    starch from, 22:699
Rosaline glasses, 21:712
Rosaniline, 21:459
Roscoelite [12271-44-2], 24:783
Rosellinia sp.
    inhibited by sorbates, 22:579
Rosemary
    extraction using SCFs, 23:466
ROSE process, 23:465
Rosin, 21:292, 459
    for fluxes, 22:496
    sampling standards for, 21:627
    in tall oil, 23:616

Rosin acid soaps
  in emulsion polymerization, *22*:999
Rosin-Rammler function, *22*:678
Rotary-blower pumps, *24*:779
Rotary kilns
  refractories for, *21*:83
Rotary-piston pumps, *24*:778
Rotational molding
  reinforced plastics, *21*:203
Rotational viscometers, *21*:384
Rotation, molecular
  spectroscopy of, *22*:636
Rotatory dispersion
  of sucrose solutions, *23*:14
Rotavirus
  vaccine for, *24*:736
Rotosil process, *21*:1038
Roughers, *21*:887
Roundworms
  veterinary drugs against, *24*:831
Rovings
  in reinforced plastics, *21*:195
Royalene
  ethylene–propylene polymer, *21*:486
RQ. See *Reportable quantity.*
RTD. See *Resonant tunneling diode.*
RTM. See *Resin-transfer molding.*
RTV. See *Room-temperature vulcanizable
  silicones.*
RTV products, *22*:119
RTV silicone rubber
  organic titanates for, *24*:329
RU-486 [*84371-65-3*], *22*:902, 903
RU-58668 [*151555-47-4*], *22*:903
Rubber
  recycling, *21*:22
  rheological measurements, *21*:427
  sampling standards for, *21*:628
  selenium in, *21*:713
  silica for, *21*:1001
  for size separation screens, *21*:908
  talc application, *23*:613
  tellurium in mfg of, *23*:805
  thiols in, *24*:28
  use of sulfur for, *23*:251
Rubber blends, *21*:578
Rubber chemicals, *21*:**460**
  sodium nitrite for, *22*:394
Rubber compounding, *21*:**481**
Rubber compounds
  reinforcement of, *21*:1025
Rubber latex, *21*:563

Rubber-modified asphalt, *21*:32
Rubber-modified PS, *22*:1025
Rubber, natural, *21*:**562**
  precipitated silica for, *21*:1025
Rubber solvent, *22*:538
Rubber, synthetic, *21*:482, 591; *22*:994
Rubella
  vaccines against, *24*:727
Rubidium [*7440-17-7*], *21*:591
  diffusion coefficient in vitreous silica,
    21:1047
Rubidium alum, *21*:596
Rubidium aluminum sulfate
    dodecahydrate [*7488-54-2*], *21*:596
Rubidium and rubidium compounds,
    21:**591**
Rubidium carbonate [*584-09-8*]
  toxicity data for, *21*:595
Rubidium chloride [*7791-11-9*]
  toxicity data for, *21*:595
Rubidium chlorostannate [*17362-92-4*],
    21:593
Rubidium hexachlorotitanate [*16902-24-
    2*], *24*:261
Rubidium hydroxide [*1310-82-3*]
  toxicity data for, *21*:595
Rubidium iodide [*7790-29-6*]
  toxicity data for, *21*:595
Rubidium monoxide [*12509-27-2*], *21*:592
Rubidium nitrate [*13126-12-0*]
  toxicity data for, *21*:595
Rubidium oxide [*18088-11-4*], *21*:591
Rubidium peroxide [*23611-30-5*], *21*:592
Rubidium–sodium alloys, *22*:347
Rubidium sulfate [*7488-54-2*]
  toxicity data for, *21*:595
Rubidium superoxide [*12137-25-6*], *21*:592
Rubidium telluride [*12210-70-7*], *21*:598
Rubidium trioxide [*12137-26-7*], *21*:592
Ruby glass
  stannous oxide in, *24*:127
Ruby red glass, *21*:712
Rule 66 requirements
  and solvents, *22*:531
Rum, *23*:8
Runoff
  of pesticides, *22*:443
Rutgers 6-12
  as mosquito repellent, *21*:241
Ruthenium, *21*:600
Rutherfordine [*12202-79-8*], *24*:643, 669
Rutherfordium, *21*:600

Rutile [1317-80-2], 24:235
  titaniums in, 24:225

S

Sabadilla, 21:601
Sabadine, 21:601
Sabaline, 21:601
Saccharides
  as sweeteners, 23:556
  in syrups, 23:583
Saccharification
  of dextrose, 23:588
Saccharin [81-07-2], 21:601; 22:401;
    23:109
  additives for refractory coatings,
    21:110
  compared to sucrose, 23:4
  reactions with aspartame, 23:559
  as sweetener, 23:564
Saccharomyces carlsbergensis
  inhibited by sorbates, 22:579
Saccharomyces cerevisiae
  inhibited by sorbates, 22:579
Saccharomyces delbrueckii
  inhibited by sorbates, 22:580
Saccharomyces fragilis
  inhibited by sorbates, 22:580
Saccharomyces lactis
  inhibited by sorbates, 22:580
Saccharomyces rouxii
  inhibited by sorbates, 22:580
Saclofen, 23:211
Safe Drinking Water Act (SDWA), 21:166;
    22:174, 532
Safeguard, 22:323
Safety, 21:601
Safety devices
  shape-memory actuators for, 21:970
Safety regulations, 24:525
Safety windshields
  PVB interlayer, 24:924
Saflex, 24:932
Safranin, 24:827
Sago
  dextrose from, 23:585
Salicyl alcohol [90-01-7], 21:601, 621
Salicylamide [65-45-2], 21:616, 617
Salicylanilide [87-17-2], 21:616, 617
Salicylic acid [69-72-7], 21:601
  as rubber chemical, 21:473

Salicylic acid and related compounds,
    21:601
Salicyloyl chloride [1441-87-8], 21:604
Salicylsalicylic acid [552-94-3], 21:614
Saligenin [90-01-7], 21:622
Salinity
  aqueous hazard ratings, 22:371
Saliretin, 21:622
Salmonella
  destruction of in foods, 22:848
Salmonella enteritidis
  inhibited by sorbates, 22:580
Salmonella heidelberg
  inhibited by sorbates, 22:580
Salmonella montevideo
  inhibited by sorbates, 22:580
Salmonella typhimurium
  inhibited by sorbates, 22:580
Salol [118-55-8], 21:613
Salsalate [532-94-3], 21:615
Salt cake, 22:403
Salt, iodized, 22:382
Samandarine [467-51-6], 22:865
Samarium, 21:626
Samarskite [1317-81-3]
  magnetic intensity, 21:876
  tantalum in, 23:660
Sampling, 21:626; 22:261
  in infrared spectroscopy, 22:638
SAMs. See Self-assembled monolayers.
SAN. See Styrene–acrylonitrile copolymer.
Sand [14808-60-7], 21:988; 22:2
  sampling standards for, 21:628
  vitreous silica from, 21:1037
Sandarac [9000-57-1], 21:297
Sandopan DTC
  surfactant, 23:493
Sandopan KST
  surfactant, 23:493
Sand reduction process, 23:736
Sand-surface ware, 21:1038
Sandvik SX, 23:388
SANECTA, 23:559
Sanforized process, 23:891
Sanitization, 22:848
α-Santalol [115-71-9], 23:871
β-Santalol [77-42-9], 23:871
Sapogenins, 22:852
Saponification, 22:299, 307
Saponins, 21:650; 22:852
  in soybeans and peanuts, 22:599
Sapphire
  for optical spectroscopy, 22:635

SARA. See *Superfund Amendments and Reauthorization Act.*
Saramet, *23*:388
Saran, *24*:882
Sarcina lutea
   inhibited by sorbates, *22*:580
Sarkosyl 0
   surfactant, *23*:493
Sarkosyl LC
   surfactant, *23*:493
Satellites
   synthetic quartz crystals for, *21*:1082
   thermoelectric power supplies for, *23*:1035
Saturated absorption
   optical spectroscopy, *22*:658
Saturated brine
   boiling point of, *23*:627
Sauflon PW, *24*:1079
Sauter mean diameter, *22*:258, 679
Savonius machine
   for wind technology, *22*:467
SAW. See *Surface acoustic wave.*
SB Latex, *22*:984
   thiols in, *24*:29
SBR. See *Styrene–butadiene rubber.*
SBR compounds, *22*:994
SBR latex
   thiols in, *24*:29
Scale
   removal by sulfamic acid, *23*:129
Scaleup, *21*:650
Scandium, *21*:650
Scanning electron microscopy, *21*:90
Scarecrows
   as bird repellents, *21*:254
SCF chromatographic separations, *23*:465
Schercamox DML
   amine oxide surfactant, *23*:525
Schercamox DMM
   amine oxide surfactant, *23*:525
Schercomid
   diethanolamine surfactant, *23*:520
Schercomid CME
   surfactant, *23*:520
Schercotaine
   surfactant, *23*:531
Schereowet
   dialkyl sulfosuccinate surfactant, *23*:498
Schiff base
   complexes of titanium, *24*:302

Schistosomiasis
   developing vaccines for, *24*:735
Schizosaccharomyces octosporus
   inhibited by sorbates, *22*:580
Schmidt number, *21*:311
Schmorlite, *24*:264
Schoepite [22972-07-2], *24*:643
Schottky barrier, *21*:708
Schottky contacts, *21*:807
Schottky diode, *21*:735
Schottky junction, *21*:1098
Schottky mechanism, *23*:673
Schugi mixer, *22*:235
Science policy
   and R&D management, *21*:274
Scientific literacy
   and R&D management, *21*:274
Scintillation crystals
   sodium iodide for, *22*:382
Scintillation detectors
   use in x-ray spectroscopy, *22*:654
Scoliosis
   shape-memory alloy devices for, *21*:974
Scotch Seal, *22*:456
Scrap
   role in steelmaking, *22*:767
Scrap rubber fuel, *21*:24
Scrap tires
   recycling rubber, *21*:23
Screening, *21*:650, 906
   in paper recycling, *21*:15
Screen printing, *21*:650
Screens
   for size separation, *21*:906
Scrubbers, *21*:650
   tail gas, *23*:379
SDWA. See *Safe Drinking Water Act.*
Seafood
   trace analysis of, *24*:515
Seafood products
   sorbates in, *22*:582
Sealants, *21*:**650**
   PVB in, *24*:935
   rheological measurements, *21*:427
Seals
   of shape-memory alloys, *21*:969
Seasonings, *21*:666
Seawater
   boiling point of, *23*:627
   composition, *22*:355
   in the Frasch process, *23*:242
   refractories from, *21*:50

rhenium in, *21*:336
specific gravity, *23*:625
sulfoxides in, *23*:217
Seaweed colloids, *21*:666
Secondary alkanesulfonates, *23*:161
Secosteroid, *22*:881
Sedimentation, *21*:**667**
  in centrifugal separation, *21*:840
  centrifuges for, *21*:849
  use in particle size measurement,
    22:268
  water treatment for steam prdn, *22*:745
Sediments
  particle size measurement for, *22*:270
SeDMC. See *Selenium*
    dimethyldithiocarbamate.
Seebeck coefficient
  for thermocouples, *23*:823
Seebeck effect, *23*:1030
Seeds
  starch from, *22*:699
Seismic control
  shape-memory alloys for, *21*:972
Selective agglomeration, *22*:249
Selectox process, *23*:445
Selenac, *21*:698, 713
Selenanthrene [262-30-6], *21*:703
Selenazofurin, *21*:713
Selenazole [288-52-8], *21*:703
Selenic acid [7783-08-6], *21*:698, 699
Selenium [7782-49-2], *21*:686; *23*:1046
  in brazing filler metals, *22*:494
  regulatory level, *21*:160
  role in steelmaking, *22*:779
Selenium amino acid chelate, *21*:697
Selenium and selenium compounds,
    21:**686**
  for self-assembled monolayers, *23*:1095
Selenium ascorbate, *21*:697
Selenium aspartate, *21*:697
Selenium citrate triturations, *21*:697
Selenium dichloride [14457-70-6], *21*:687,
    697, 699
Selenium diethyldithiocarbamate [5456-
    28-7], *21*:698
  for rubber, *21*:713
Selenium dimethyldithiocarbamate
    (SeDMC) [144-34-3], *21*:466
Selenium dioxide [7446-08-4], *21*:688,
    698, 699
Selenium disulfide [7488-56-4], *21*:698
Selenium hexafluoride [7783-79-1], *21*:699

Selenium hypofluorite [27218-12-8],
    21:699
Selenium lysinate, *21*:697
Selenium monobromide [7789-52-8],
    21:699
Selenium monochloride [10025-68-0],
    21:699
Selenium monoxide [12640-89-0], *21*:701
Selenium oxybromide [7789-51-7], *21*:699
Selenium oxychloride [7791-23-3], *21*:698,
    699
Selenium pentoxide [12293-89-9], *21*:701
Selenium sulfide bentonite, *21*:697
Selenium–Tellurium Development
    Association, *21*:715
Selenium tetrabromide [7789-65-3], *21*:699
Selenium tetrachloride [10026-03-6],
    21:698, 699
Selenium tetrafluoride [13465-66-2],
    21:699
Selenium trioxide [13768-86-0], *21*:699
Selenocyanogen [27151-67-3], *21*:702
Selenocysteine, *21*:702
Selenomycin, *21*:713
Selenophene [288-05-1], *21*:703
Selenotifen, *21*:713
Selenourea [630-10-4], *21*:698, 703
Selenous acid [7783-00-8], *21*:698, 699
1,4-Selenoxane [5368-46-7], *21*:703
Selexol process, *23*:440
Self-assembled monolayers (SAMs),
    23:1087
Sellmeier dispersion, *21*:1057
Selsun Blue, *21*:713
Semen
  preservative for, *23*:113
Semiconductor diodes
  for optical spectroscopy, *22*:635
Semiconductor materials
  thin films for, *23*:1066
Semiconductors
  amorphous, *21*:**750**
  cleaning using SCF, *23*:465
  compound, *21*:**763**
  grown in microgravity, *22*:623
  metal tellurides for, *23*:789, 805
  polishing agents for, *21*:1000
  pure silicon for, *21*:1094
  role in thermoelectric energy
    conversion, *23*:1032
  for sensors, *21*:820
  silicon-based, *21*:**720**

silicon halides in prdn of, *22*:36
thin-film metallization for, *23*:1056
titanium coatings, *24*:323
ultrapure water for, *21*:326
vitreous silica in, *21*:1067
Semperfresh, *23*:7
Sensors, *21*:**816**
  for hydrogen sulfide detection, *23*:282
  thin films for, *23*:1087
  titanium dioxide in, *24*:238
  in trace analysis, *24*:510
  use of atomic fluorescence spectroscopy,
    22:653
  vitreous silica fibers in, *21*:1069
Separation
  centrifugal, *21*:**828**
  liquid–liquid, *21*:842
  low energy, *21*:923
  magnetic, *21*:**876**
  reverse micelle, *23*:465
  reverse osmosis membranes for, *21*:304
  size, *21*:**901**
  of steam from water, *22*:737
  of sugar alcohols, *23*:103
  use of rubidium compounds, *21*:598
  use of sulfolane, *23*:139
Separation cell, *23*:732
Separations process synthesis, *21*:**923**
Separatrix, *21*:929
Sep-Pak, *24*:493
Sequestering agents
  for actinides, *24*:79
  lignosulfates as, *23*:205
SERC. See *State Emergency Response
    Commission.*
Sericin
  in silk, *22*:156
Sericin gum, *23*:549
Serine
  in silk, *22*:156
  in soybeans and other oilseeds, *22*:596
Serology, *21*:962
Serotonin [*50-67-9*], *22*:939
  reuptake inhibitors, *22*:944
Serpentine
  magnetic intensity, *21*:876
Serratia marcescens
  inhibited by sorbates, *22*:580
Sers. See *Surface-enhanced Raman
    scattering.*
Sertraline [*79617-96-2*], *22*:944
Sesquiterpene, *23*:833

Sesterterpene, *23*:833
SET. See *Single-electron transistor.*
Sethoxydim
  thiols in, *24*:29
Seville orange
  sweetener from, *23*:571
Sewage
  dewatering of, *21*:863
  hydrothermal oxidation of, *23*:470
  tanks for, *23*:644
Sewage gases
  sulfur removal from, *23*:444
Sfc. See *Supercritical fluid
    chromatography.*
SFR
  styrene inhibitor, *22*:973
Shade stain, *22*:697
Shale oil, *21*:962
Shampoo
  sorbitan fatty esters in, *23*:111
  sulfonates in, *23*:205
Shape control
  shape-memory alloys for, *21*:972
Shape factors
  role in size measurement, *22*:257
Shape-memory alloys, *21*:**962**
Sharpless catalyst, *24*:301
Shaving creams
  amine soaps in, *22*:324
Shearometer, *21*:423
Shear stresses, *21*:354
Shear viscosity, *21*:364
Sheet
  of polystyrene, *22*:1065
Sheet molding compound (SMC), *21*:199
Sheet rubber, *21*:564
Shellac [*9000-59-3*], *21*:299
  sampling standards for, *21*:628
Shellolic acid, *21*:300
Sherwood number, *21*:310
Shield, *22*:323
Shigella
  vaccine against, *24*:742
Shinetsu P-10, *21*:1034
Shingles, *21*:450
Shipping, *24*:531
Ships
  steam turbines for, *22*:755
Shoe products
  SBR use, *22*:1011
Shoe soles
  precipitated silica in, *21*:1025

Shortening, *21*:976
  emulsifiers, *23*:110
  soybeans and other oilseeds in, *22*:610
Short-rotation woody crop (SRWC), *21*:227
Shortstopping agents
  for SBR, *22*:1002
Shredded tire chips, *21*:27
Shredding, *22*:279
Shrimp meal, *21*:976
Sialon
  as tool materials, *24*:430
SIC. See *Standard Industrial
  Classification Code.*
Siderite, *21*:976
  magnetic intensity, *21*:876
  rhenium in, *21*:336
Siding
  PVC use, *24*:1040
Siemens process, *22*:41
  for pure silicon, *21*:1092
Sienna, burnt, *21*:976
Sieves, *21*:976
  role in particle size measurement,
    22:264
Signaling smokes, *21*:976
Significant new use rule (SNUR), *21*:166
Sila-crowns, *22*:70
Silane, *21*:976, 1087
  for chemical vapor deposition, *23*:1060
  use in silicon purification, *21*:1092
Silane adhesion promoters, *21*:656
Silanediols
  titanates of, *24*:289
Silanes, *22*:**38**
Silanol, *21*:976
  vibrational peaks of, *22*:516
Silanol polycondensation, *22*:89
Silatranes, *22*:70
Silica [*7631-86-9*], *21*:**977**, 119, 1087; *22*:3
  amorphous, *21*:**1005**
  in brazing filler metals, *22*:492
  as catalyst support, *23*:389
  refractories from, *21*:49
  as refractory material, *21*:53, 95
  as release agent, *21*:210
  SCF solution of, *23*:472
  solubility in steam, *22*:728, 730
  synthetic quartz crystals, *21*:**1076**
  Type V gel, *22*:519
  vitreous, *21*:**1032**
Silica–alumina [*37287-16-4*], *24*:862
Silica, brick, *21*:1084

Silica–calcium–phosphorus pentoxide
  sol–gel hybrids of, *22*:527
Silica fibers, *21*:123; *22*:639
Silica, fumed
  critical surface tension of, *22*:148
Silica gels, *21*:993, 1009, 1020
  pK_a of silanol groups, *22*:9
  silylating agent on, *22*:149
Silica glass
  for optical spectroscopy, *22*:635
Silica M, *21*:1060
Silica–PDMS
  from sol–gel technology, *22*:525
Silica–polydimethylsiloxane
  from sol–gel technology, *22*:525
Silica sand, *23*:734
Silica sols, *22*:25
Silicate grouts, *22*:453
Silicates, *21*:1084; *22*:**1**
  as release agent, *21*:210
Silica W, *21*:988
Silicic acid [*1343-98-2*]
  in grouts, *22*:455
Silicides, *21*:1084, 1111
  as refractory coatings, *21*:103
Silicomanganese, *21*:1118
Silicon [*7440-21-3*], *21*:1084
  in brazing filler metals, *22*:484, 487
  in ferrous shape-memory alloys, *21*:965
  in pig iron, *22*:767
  semiconductors for sensors, *21*:820
  sensors based on, *21*:820
  in shape-memory alloys, *21*:964
  in steel, *22*:775
  thin films of, *23*:1060
  in tool steels, *24*:398
Silicon–aluminum alloys
  use of sodium with, *22*:346
Silicon and silicon alloys, *21*:**1084**
  chemical and metallurgical, *21*:**1104**
  pure silicon, *21*:**1084**
Silicon-based semiconductor industry,
  21:720
Silicon carbide [*409-21-2*], *21*:56, 119,
  1109
  as refractory coating, *21*:105
  as refractory material, *21*:57
  thin films of, *23*:1061
Silicon carbide fibers, *21*:122
Silicon chips
  silver alloy solder for, *22*:174
Silicon compounds
  silanes, *22*:**38**

silicones, *22*:**82**
silicon esters, *22*:**69**
silicon halides, *22*:**31**
silylating agents, *22*:**143**
synthetic inorganic silicates, *22*:1
Silicon dioxide [*7631-86-9*], *21*:977, 1005, 1076
    as refractory material, *21*:57
    salt additive, *22*:368
    silicon from, *21*:1104
Silicon esters, *22*:**69**
Silicon germanium
    thermoelectric material, *23*:1034
Silicon halides, *22*:**31**
Silicon halohydrides, *22*:31
Siliconized acrylic sealants, *21*:656
Silicon metal
    silica from, *21*:1019
    trichlorosilane in mfg of, *22*:36
Silicon monoxide, *21*:1087; *23*:1046
    as refractory coating, *21*:113
Silicon nitride [*12033-89-5*], *21*:58, 119, 1109
    films as refractory coatings, *21*:98
    thin films of, *23*:1061, 1064
    as tool materials, *24*:427
Silicon-29 nmr
    for hydrolysis in sol–gel technology, *22*:505
Silicon oxide
    substrate for self-assembled monolayer, *23*:1089
Silicon oxynitride
    thin films of, *23*:1065
Silicon, polycrystalline
    from silanes, *22*:46
Silicon Technology Competitive
    Cooperative, *21*:1107
Silicon tetrachloride, *22*:31
    silica from, *21*:1019
    use in silicon purification, *21*:1092
    vitreous silica from, *21*:992, 1037
Silicon tetrafluoride
    precipitated silica from, *21*:998
    pyrogenic silica from, *21*:1026
Silicone fluids, *22*:101
Silicone foamed rubber, *22*:113
Silicone latex sealants, *21*:656
Silicone LIM rubber, *22*:112
Silicone networks, *22*:95
Silicone oils, *22*:107
    for precipitated silica, *21*:997

Silicone polymers
    fumed silica reinforcement for, *21*:1027
Silicone PSAs, *22*:116
Silicone resins, *22*:113
Silicone rubber, *22*:109
    silicon esters for, *22*:78
Silicones, *22*:**82**
    alkylsilanes in prdn of, *22*:34
    cured by titanates, *24*:323
    fractionation using SCFs, *23*:469
    as release agents, *21*:210
    silicon for, *21*:1108
Silicone sealants, *21*:654
Silicosis, *21*:999, 1120
Silk, *22*:**155**
    as sutures, *23*:542
    titanated inks for, *24*:329
Silkworm
    source of silk, *22*:155
Silky II Polydek, *23*:543
Sillimanite [*12141-45-6*], *21*:53; *22*:163
    refractories from, *21*:49
Siloxane polymers
    by emulsion polymerization, *22*:93
Siloxanes, *21*:1087
Siloxene [*27233-73-4*], *22*:39, 44
Silver [*7440-22-4*], *22*:163
    in brazing filler metals, *22*:492
    complex thiosulfates, *24*:55
    containers for quartz synthesis, *21*:1080
    emissivities of, *23*:827
    extraction of, *22*:382
    plating with methanesulfonic acid, *23*:326
    regulatory level, *21*:160
    single dc-diode sputtering of, *23*:1053
    in solders, *22*:485
    tellurium in ores, *23*:782
    vapor pressure of, *23*:1046
Silver acetylide [*7659-31-6*], *22*:180
Silver and silver alloys, *22*:**163**
    as brazing filler metals, *22*:493
Silver antimony germanium telluride
    thermoelectric material, *23*:1034
Silver arsenite [*15122-57-3*], *22*:179
Silver behenate [*2489-05-6*], *22*:183
    for photographic paper, *22*:192
Silver–cadmium alloys
    as shape-memory alloys, *21*:964
Silver compounds, *22*:**179**
Silver(I) compounds, *22*:180

Silver dichromate [7784-02-3], 22:182
Silver diethyldithiocarbamate [1470-61-7], 22:190
Silver hydrogen sulfate [19287-89-9], 22:184
Silver laurate [18268-45-6], 22:183
   for photographic paper, 22:192
Silver molybdate [13765-74-7], 22:192
Silver nitrate [7761-88-8]
   silver compounds, 22:179
Silver oxalate [533-51-7], 22:183
Silver(II) oxide [1301-96-8, 35366-11-1], 22:186
Silver oxide–zinc cell, 22:190
Silver perbromate [54494-97-2], 22:184
Silver perchlorate [7783-93-9]
   in organic solvents, 22:184
Silver periodate [15606-77-6], 22:184
Silver permanganate [7783-98-4], 22:184
Silver phosphate [7784-09-0], 22:184
Silver pyrophosphate [13465-97-9], 22:184
Silver selenide [1302-09-6], 22:179
Silver stearate [3507-99-1], 22:183
   for photographic paper, 22:192
Silver subfluoride [1302-01-8], 22:182
Silver sulfadiazine [22199-08-2], 22:191
Silver sulfonantimonite [15983-65-0], 22:179
Silver telluride [12653-91-7], 22:179
Silver tetrafluoroborate [14104-20-2], 22:184
Silver–thallium alloys, 23:953
Silver thiosulfate [23149-52-2], 22:185
Silylating agent, 22:143
Silyl ethers
   of sucrose, 23:65
Simazine [122-34-9], 22:425; 24:510
Simultaneous heat and mass transfer, 22:195
Single crystals, 22:222
Single-electron transistor (SET), 21:787
Single-ply roofing
   as roofing material, 21:443
Single radial immunodiffusion, 24:734
Sintering, 22:242
Sintering ores
   particle-bonding mechanism, 22:224
SIP. See State implementation plan.
Siroc method
   silicate grouting, 22:453
Sisko model, 21:350
η-Sitosterol [83-46-5], 22:863
   in soybeans and other oilseeds, 22:597

SI Units, 24:629
Size enlargement, 22:222
Size measurement of particles, 22:256
Size reduction, 22:279
Sizes, 21:296
   starch in, 22:712
Skin, 22:296
Skin-care products
   titanium dioxide for, 24:239
Skin-So-Soft
   as mosquito repellent, 21:244
Skutterudites
   thermoelectric material, 23:1034
Slagceram, 22:296
Slags
   effect on refractories, 21:81
Slated sulfur, 23:257
S'Lec, 24:932
Sliding plate rheometer, 21:402
Slimes
   tellurium from, 23:786
Slimicides, 22:296
Slip agents, 21:207
Sludge handling
   in paper mills, 21:19
Slurry fuels
   rheological measurements, 21:427
Small-angle x-ray scattering
   for sol–gel technology studies, 22:507
Smallpox vaccine, 24:727
SMART, 22:979
Smart material
   shape-memory alloys for, 21:972
Smart sensor, 21:821
SMC. See Sheet molding compound.
SME. See Sucrose monoesters.
Smelting
   sulfuric acid as by-product, 23:378
Smog, 22:296
Smokes, 22:296
SMR. See Standard Malaysian Rubber.
SNAP-9A, 23:1035
SNOX, 23:449
SNUR. See Significant new use rule.
Soap, 22:297
   centrifugal separation of, 21:859
   in emulsion polymerization, 22:999
   sampling standards for, 21:628
   silicates in, 22:21
   strontium compounds in, 22:954
   sulfonates in, 23:205
   sulfonic acid-derived dyes in, 23:206

as synthetic surfactant, 23:478
triarylmethane dyes for, 24:565
use of sodium sulfate for, 22:410
use of sulfur for, 23:251
Soapstone, 23:611
Social concerns
    and research/technology management,
    21:268
SOCMI. See *Synthetic organic chemical
    industry.*
Soda, 22:326
Sodalite
    silylating agents for, 22:147
Sodamide, 22:332
Soddyite [12196-99-5], 24:643
Soderberg electrodes
    in ferrosilicon prdn, 21:1114
Sodium [7440-23-5], 22:327; 24:854
    complexes with sugar alcohols, 23:103
    diffusion coefficient in vitreous silica,
        21:1047
    for reduction of tantalum, 23:665
    from sodium chloride, 22:374
    in steam, 22:737
    vitreous silica impurity, 21:1036
Sodium acetylide, 22:332
Sodium aluminate [1302-42-7]
    in silicate grouts, 22:453
Sodium amalgam, 22:339
Sodium and sodium alloys, 22:**327**
Sodium anthracene-9-carboxylate
    surface-active agents, 22:332
Sodium anthraquinone disulfonate
    use in sulfur recovery, 23:444
Sodium atoms
    analysis using Raman spectroscopy,
        22:651
Sodium bicarbonate [144-55-8]
    as blowing agent, 21:479
    centrifugal separation of, 21:867
Sodium bisulfate [7681-38-1], 22:403
Sodium bisulfite [7631-90-5], 23:314;
    24:856
    effect on reverse osmosis membranes,
        21:315
Sodium bromate [7789-38-0]
    sodium bromide from, 22:378
Sodium bromide [7647-15-6], 22:**377**
Sodium carbonate [497-19-8], 22:2, 354,
    378
    in sodium chloride prdn, 22:361
    use in sulfur recovery, 23:444

Sodium carbonyl, 22:332
Sodium chlorate [7775-09-9], 21:695;
    23:197
    from sodium chloride, 22:374
Sodium chloride [7647-14-5], 22:**354**;
    24:858
    centrifugal separation of, 21:867
    controlled release, 22:374
    in sodium nitrate, 22:390
    solubility in steam, 22:728, 730
Sodium β-chloroethanesulfonate [15484-
    44-3], 24:856
Sodium N-chlorosulfamate [13637-90-6],
    23:123
Sodium cocoyl isothionate, 22:322
Sodium compounds, 22:**354**
    sodium bromide, 22:**377**
    sodium chloride, 22:**354**
    sodium halides, 22:**354**
    sodium iodide, 22:**380**
    sodium nitrate, 22:**383**
    sodium nitrite, 22:**394**
    sodium sulfate, 22:**403**
    sodium sulfides, 22:**411**
Sodium cyanide
    assay for tin, 24:114
Sodium cyclamate [139-05-9], 23:566
Sodium cyclohexylsulfamate [139-05-9],
    23:172
    from sulfamic acid, 23:130
Sodium di-n-butyldithiocarbamate
    (NaDBC) [136-30-1], 21:466
Sodium dichloromethanesulfinate
    [36829-83-1], 23:273
Sodium N,N-dichlorosulfamate [13637-
    67-7], 23:123
Sodium dihydrogen monosilicate
    heptahydrate [27121-04-6], 22:5
Sodium dihydrogen monosilicate
    octahydrate [13517-24-3], 22:5
Sodium dihydrogen monosilicate
    pentahydrate [35064-64-3], 22:5
Sodium dihydrogen monosilicate
    tetrahydrate [10213-79-3], 22:5
Sodium dithionate
    for cleaning membranes, 21:317
Sodium dithionite [7775-14-6], 23:316
Sodium dititanate [12164-19-1], 24:251
Sodium ethyl thiosulfate [26264-37-9],
    24:65
Sodium fatty acid ester sulfonates, 23:162
Sodium ferrocyanide
    in dendritic salt prdn, 22:363

Sodium ferrocyanide decahydrate
  salt additive, 22:368
Sodium formaldehyde sulfoxylate
  [149-44-0], 23:319
Sodium–gold alloy, 22:347
Sodium halides, 22:**354**
Sodium hydrosulfide [16721-80-5], 22:412;
  23:280
  strontium carbonate by-product, 22:953
Sodium hydrotelluride [23624-18-2],
  23:794
Sodium hydroxide [1310-73-2], 22:419
  sodium bromide from, 22:378
  in sodium chloride prdn, 22:361
  solubility in steam, 22:728
  water additive of steam prdn, 22:740
Sodium hypochlorite
  in paper recycling, 21:18
Sodium iodide [7681-82-5], 22:**380**; 23:961
Sodium N-lauroylsarcosinate [7631-98-3],
  23:493
Sodium–lead alloys, 22:347
Sodium levothyroxine [55-03-8], 24:102
Sodium metabisulfite, 23:315
Sodium metabisulfite [7681-57-4], 23:311
Sodium metasilicate [1344-09-8], 22:4
Sodium metatitanate [12034-34-3], 24:251
Sodium metavanadate [13718-26-8],
  24:800
Sodium monophosphate
  solubility in steam, 22:728
Sodium monopolysilicate [6834-92-0], 22:5
Sodium naphthalene complex, 22:331
Sodium nitrate [7631-99-4], 22:**383**
Sodium nitrite [7362-00-0], 22:**394**
  in sodium nitrate, 22:390
Sodium olefin sulfonate, 22:322
Sodium N-oleoyl-N-methyltaurate
  [137-20-2], 23:499
Sodium orthophosphate
  water additive of steam prdn, 22:740
Sodium orthosilicates, 22:15
Sodium orthovanadate [13721-39-6],
  24:800
Sodium pentafluorostannite [22578-17-2],
  24:130
Sodium pentatitanate [12034-52-5], 24:251
Sodium phenanthrene-9-carboxylate
  surface-active agents, 22:332
Sodium phosphate
  boiler water additive, 22:738
  solubility in steam, 22:728

Sodium phosphide, 22:332
Sodium poly(acrylate)
  in de-inking, 21:14
Sodium–potassium alloy, 22:347
Sodium pyrosulfite [7681-57-4], 23:313
Sodium pyrovanadate [13517-26-5], 24:800
Sodium saccharin [128-44-9]
  as sweetener, 23:565
Sodium salicylate [54-21-7], 21:603, 612
Sodium salt, 23:203
Sodium selenate [13410-01-0], 21:687, 698
Sodium selenide [1313-85-5], 21:698
Sodium selenite [10102-18-8], 21:687, 698
Sodium selenocyanate [4768-87-0], 21:701
Sodium selenosulfate [25468-09-1], 21:697,
  702
Sodium sesquisilicate [1344-09-8], 22:17
Sodium silicate [1344-09-8], 22:1, 419
  in chemical grouts, 22:453
  in pulping, 21:14
  silica for, 21:999
  silica from, 21:1019
  silylating agents for, 22:147
Sodium silicide, 22:332
Sodium silicoaluminate
  salt additive, 22:368
Sodium sorbate [7757-81-5], 22:574
Sodium stannate [12058-66-1], 24:128
Sodium stearate [822-16-2], 22:324
Sodium sulfate [7757-82-6], 22:**403**, 403;
  23:312; 24:496
  from sodium chloride, 22:374
  in sodium nitrate, 22:390
  solubility in steam, 22:728
Sodium sulfate decahydrate [7727-73-3],
  22:403
Sodium sulfide [1313-82-2], 22:**411**;
  23:312; 24:854
Sodium sulfide nonahydrate [1313-84-4],
  22:415
Sodium sulfite [7757-83-7], 23:311, 312
  sulfur dioxide absorption, 23:447
Sodium sulfites, 22:419
  oxygen scavenger for steam, 22:742
  in steam, 22:737
  water additive of steam prdn, 22:740
Sodium tellurate [10101-25-8], 23:788
Sodium tetraborate
  in sodium nitrate, 22:390
Sodium tetrasulfide [12034-39-8], 22:412;
  23:284
Sodium thiocyanate [540-72-7], 23:322

Sodium thioglycolate [*367-51-1*], *24*:6
Sodium thiosulfate [*7772-98-7*], *24*:53
  use in chlorine analysis, *21*:644
Sodium thiosulfate pentahydrate
    [*10102-17-7*], *24*:53
Sodium titanate [*12034-34-3*], *24*:235, 302
Sodium tripolyphosphate, *22*:419
Sodium trititanate [*12034-36-5*], *24*:251
Sodium uranate [*13721-31-4*], *24*:650
Sodium vanadate
  use in sulfur recovery, *23*:444
Sodyesul Black 4GCF
  sulfur dye, *23*:343
Sofsilk, *23*:543
Softcon, *24*:1079
Soft drinks
  aspartame in, *23*:559
Softeners, *21*:527
  sorbitol solutions as, *23*:108
Softsoap, *22*:323
Soft sugars, *23*:36
Soil chemistry of pesticides, *22*:**419**
Soil conditioners, *22*:452
Soil conditioning
  for agriculture, *22*:457
Soil disinfectant
  sodium bisulfate for, *22*:411
Soil remediation, *22*:252
Soils
  cleaning using SCF, *23*:465
  particle size measurement for, *22*:270
  sampling standards for, *21*:628
  trace analysis of, *24*:518
Soil stabilization, *22*:**452**
  sodium silicates in, *22*:24
Soiltex, *22*:459
Soil vapor monitoring, *23*:656
Solar cells, *21*:1094; *23*:1062
  pure silicon, *21*:1101
  refractory coatings for, *21*:113
  tellurium in, *23*:791
  thin films for, *23*:1066
  titanium coatings, *24*:323
  vitreous silica coatings for, *21*:1069
Solar energy, *22*:**465**, 393, 410
  captured in sugar, *23*:79
  silver mirrors for, *22*:176
Solar panels
  amorphous semiconductors in, *21*:750
Solar salt, *22*:354
Solar surface
  analysis by γ-ray spectroscopy, *22*:656

Solasodine [*126-17-0*], *22*:863
Soldering, *22*:482
Soldering alloys
  silver in, *22*:174
Solders
  thin films of, *23*:1071
  tin alloys as, *24*:117
  tin compounds in, *24*:129
  use of selenium in, *21*:711
Solders and brazing filler metals, *22*:**482**
Sol–gel technology, *22*:**497**
  hydrolysis of tetraalkoxysilanes, *22*:72
  for silica gels, *21*:996
  silica gels as, *21*:1023
  of silicic acid esters, *22*:9
  for thin films, *23*:1073
  titanates for, *24*:332
  for vitreous silica, *21*:1037
Solid bridges
  binder, *22*:224
Solidification, *22*:245
Solid-phase microextraction, *24*:493
Solids disposal, *23*:734
Solid tantalum capacitors, *23*:670
Solid waste disposal
  regulation of, *21*:159
Sols
  of alumina, *22*:521
  of silica, *21*:993, 1009
Solubility
  retrograde, *23*:457
Solubility parameter
  role for a supercritical fluid, *23*:454
Solubility products
  of silver compounds, *22*:181
Solubilized S Black 1 [*1326-83-6*], *23*:343
Solution SBR, *22*:995
Solution theory, *22*:534
140 Solvent, *22*:538
Solvent-based coatings, *22*:568
Solvent Blue 23 [*2152-64-9*], *24*:556
Solvent dyes
  in wood stains, *22*:692
Solvent extraction, *22*:570; *23*:729
  of caffeine from tea, *23*:762
  use in tantalum recovery, *23*:661
Solvent-release acrylics
  as sealants, *21*:660
Solvents
  for BTX extraction, *23*:137
  sampling standards for, *21*:628
  steam as, *22*:727

Solvents, industrial, 22:**529**
  DMSO as, 23:228
Solvent vapors
  sensors for, 21:824
Somatotropins, 22:571
Sonic atomizer, 22:673
Sonic devices
  as repellents, 21:260
Sorbates, 22:571
Sorbic acid [110-44-1], 22:**571**
Sorbic acid anhydride [13390-06-2], 22:575
1,5-Sorbitan, 23:96
Sorbite, 22:590
Sorbitol [50-70-4], 22:590; 23:94, 585
  salt additive, 22:368
  as sweeteners, 23:556
SORBO, 23:105
L-Sorbose
  from sorbitol, 23:102
Sorboyl chloride [2614-88-2], 22:575
Sorghum, 22:590
  dextrose from, 23:585
  protection from birds, 21:254
Sorption balance, 21:1066
Sorption isotherms
  for pesticides, 22:440
Sorting
  unit operation of tea processing, 23:757
Sour gases
  hydrogen sulfide in, 23:275
  sulfur removal from, 23:436
Soxhlet's modification
  of reducing sugar tests, 23:15
Soya protein
  centrifugal separation of, 21:860
Soybean oil
  in soap manufacture, 22:302
Soybeans
  extraction using SCFs, 23:467
Soybeans and other oilseeds, 22:**591**
SoyDiesel, 22:607, 612
Soymilks, 22:615
Space chemistry, 22:619
Spacecraft
  power supplies, 23:1034
Spacecraft materials
  action of vacuum, 24:758
Space probes
  thermoelectric power supplies for,
    23:1035
Space processing, 22:**620**
  synthetic quartz crystals for, 21:1082

Space shuttle
  microgravity experiments, 22:621
  use of silver, 22:168
Space shuttle vehicles
  vitreous silica windows, 21:1068
Space technology
  fused silica, 21:1000
Space vehicles
  refractory coatings for, 21:106
Span 20
  surfactant, 23:515
Span 40
  surfactant, 23:515
Span 60
  surfactant, 23:515
Span 65
  surfactant, 23:515
Span 80
  surfactant, 23:515
Span 85
  surfactant, 23:515
Spandex, 22:627
Spandex fibers, 24:696
SPE. See Sucrose polyester.
Spearmint oil
  sulfoxides in, 23:217
Specialty chemicals
  analysis using x-ray fluorescence,
    22:655
Specialty elastomers, 21:488
Specialty soaps, 22:321
Specific breakage rate
  particle size reduction, 22:281
Specific rotation
  of sugar solutions, 23:13
Specific viscosity, 21:356
Spectrofluorometers, 22:652
Spectroscopy
  time-resolved, 22:643
Spectroscopy, optical, 22:**627**
  determination of silver, 22:187
Spectrosil, 21:1034, 1049, 1058
Spectrosil WF, 21:1034
Spersal, 22:459
Sphalerite [12169-28-7]
  sulfide ore, 23:244
  tellurium in, 23:782
Sphene, 24:264
Spider
  source of silk, 22:155
Spiegler-Kedem
  reverse osmosis models, 21:308

Spill prevention control and
  countermeasure plan, *21*:166
Spindle
  in sugar density measurement, *23*:15
Spinel [*1302-67-6*], *21*:56
  as refractory material, *21*:57
Spin-on-glass, *23*:1074
Spiral jet mill, *22*:291
Spirit, *22*:323
Spirit blue [*2152-64-9*], *24*:556
Spironolactone [*52-01-7*], *22*:886
Spirorenone [*74220-07-8*], *22*:887
SPLENDA, *23*:569
Sporobolomyces sp.
  inhibited by sorbates, *22*:580
Sporotrichum pruinosum
  inhibited by sorbates, *22*:579
Spray drying
  for instant tea, *23*:761
Spray painting
  use of SCFs, *23*:465
Spray pyrolysis
  thin films by, *23*:1073
Spray recrystallization
  of protein powders, *23*:473
Sprays, *22*:**670**
Spray-up process, *21*:196
Spreading
  in surfactants, *23*:485
Sputtering, *21*:94; *22*:691
  thin films from, *23*:1049
SQ-27239 [*85197-77-9*], *22*:906
Squalamine [*148717-90-2*], *22*:870
Squalane [*111-01-3*], *23*:874
Squalane–carbon dioxide
  enhancement factor for, *23*:463
Squalene [*111-02-4*], *23*:874
Srgs. See *Stimulated Raman gain
  spectroscopy.*
SRI oxime V [*59691-20-2*]
  potential sweetener, *23*:573
Srs. See *Stimulated Raman scattering.*
SRWC. See *Short-rotation woody crop.*
SRWC fuel, *21*:227
Stabileze, *24*:1064
Stabilizers
  organotins as, *24*:145
  for poly(vinyl chloride), *24*:131
  sorbitol as, *23*:112
  starches as, *22*:714
  sulfur polymer cement as, *23*:262
Stachyose
  in soybeans and other oilseeds, *22*:598

Staebler-Wronski effect, *21*:762
Stagonospora sp.
  inhibited by sorbates, *22*:579
Stainless steel, *22*:781, 817
  austenitic, *21*:845
  brazing of, *22*:484
  dielectric constants of, *22*:761
  ferrosilicon in prdn of, *21*:1115
  as sampling probe, *21*:634
  screens for size separation, *21*:909
  selenium in prdn of, *21*:710
  for suture needles, *23*:551
  for tanks and pressure vessels, *23*:643
  use with steam, *22*:761
304 Stainless steel
  corrosion rates in acid, *23*:122
Stains, industrial, *22*:**692**
Standamox O1
  amine oxide surfactant, *23*:525
Standapol
  alcohol surfactant, *23*:502
Standapol A
  surfactant, *23*:501
Standard Industrial Classification Code
  (SIC), *21*:166
Standard Malaysian Rubber (SMR),
  21:565
Standard platinum resistance
  thermometer, *23*:815
Standard Revertex, *21*:582
Stannane [*2406-52-2*], *24*:135
Stannasol A and B, *24*:128
Stannic arsenate [*35568-59-3*], *24*:129
Stannic bromide [*7789-67-5*], *24*:124
Stannic chloride [*7646-78-8*], *24*:855
  use in rubidium purification, *21*:593
Stannic chloride pentahydrate [*10026-
  06-9*], *24*:126
Stannic iodide [*7790-47-8*], *24*:124
Stannic molybdate [*34782-17-7*], *24*:129
Stannic phosphate [*15142-98-0*], *24*:129
Stannic sulfide [*1315-01-1*], *24*:129
Stannic vanadate [*66188-22-5*], *24*:129
Stannite [*12019-29-3*], *24*:123
Stannous acetate [*638-39-1*], *24*:149
Stannous bromide [*10031-24-0*], *24*:124
Stannous chloride [*7772-99-8*], *24*:130
Stannous chloride dihydrate [*10025-69-1*],
  24:124
Stannous ethylene glycoxide [*68921-71-1*],
  24:130, 149
Stannous 2-ethylhexanoate [*301-10-0*],
  24:130, 149

Stannous fluoride [7783-47-3], 24:126, 130
Stannous formate [2879-85-8], 24:149
Stannous gluconate [35984-19-1], 24:149
Stannous iodide [10294-70-9], 24:124
Stannous oleate [1912-84-1], 24:149
Stannous oxalate [814-94-8], 24:130, 149
Stannous oxide [21651-19-4], 24:130
Stannous oxide hydrate [12026-24-3],
    24:127
Stannous pyrophosphate [15578-26-4],
    24:129, 130
Stannous stearate [6994-59-8], 24:149
Stannous sulfide [1314-95-0], 24:129
Stannous tartrate [815-85-0], 24:130, 149
Stanoamul
    ethoxylated alcohol surfactant, 23:508
Staphylococci
    antimicrobial agent for, 24:828
Staphylococcus
    destruction of in foods, 22:848
Staphylococcus aureus
    inhibited by sorbates, 22:580
Starch [9005-25-8], 22:**699**
    centrifugal separation of, 21:860
    corn sweeteners from, 23:582
    sorbitol from, 23:105
    sulfation of, 23:171
Starch acetates, 22:715
Starch phosphate monoesters, 22:711
Starch pyrodextrins, 22:713
Starch syrup, 23:597
Stark spectra, 22:636
State Emergency Response Commission
    (SERC), 21:166
State implementation plan (SIP), 21:166
Statistical association fluid theory
    for supercritical fluids, 23:463
Staurolite
    magnetic intensity, 21:876
Staverman reflection coefficient
    role in reverse osmosis, 21:308
Steam [7732-18-5], 22:**719**; 23:728
    for processing soybeans and other
        oilseeds, 22:601
    refractories for, 21:79
    sampling standards for, 21:628
    solubility of silica in, 21:983
    for sterilization, 22:839
    sulfuric acid by-product, 23:379
Steam ejectors, 24:777
Steam generator, 22:733
Steam irons
    sulfamic acid cleaner for, 23:129

Steam reforming
    use of steam, 22:757
Steam turbines, 22:721
    steel for, 22:822
Stearato chromic chloride [12768-56-8],
    21:214
Stearic acid [57-11-4], 22:764; 23:617
    for monolayer formation, 23:1077
    as rubber chemical, 21:473
    separation coefficients for, 21:926
    surfactants from, 23:528
Stearic acid monolayer [57-11-4]
    release substrate, 21:212
Stearylamine
    thin films of, 23:1086
Steatite, 22:764
Steel [52013-36-2], 22:**765**; 23:551. (See
    also *Stainless steel.*)
    for centrifuges, 21:845
    dielectric constants of, 22:761
    emissivities of, 23:827
    ferrosilicon in prdn of, 21:1114
    mechanical properties of, 22:160
    Rockwell hardness of, 24:397
    sampling in mills, 21:641
    selenium in prdn of, 21:710
    for steam turbines, 22:748
    sulfur in prdn of, 23:260
    as sutures, 23:542
    tellurium machinability additives,
        23:799
    as tool materials, 24:398
    vanadium in, 24:794 *H* steels, 22:801
Steel tire cord, 24:169
Steep roofing, 21:450
Stefan-Boltzman law, 23:825
Steffen process, 23:603
Steinhart-Hart equation, 23:821
Stellite, 23:281
    for centrifuges, 21:847
Stellite tools, 24:404
Stents
    of shape-memory alloys, 21:974
Steol, 23:502
    alcohol surfactant, 23:502
Stepanol AM
    surfactant, 23:501
Stepanol MG
    surfactant, 23:501
Stepanol WA 100
    surfactant, 23:501
Stepanol WAT
    surfactant, 23:501

Stephanite [1302-12-1], 22:170
Sterculic acid, 22:609
  in cottonseed, 22:598
Sterile filling
  for sterilized materials, 22:846
Sterilization techniques, 22:**832**
  for sutures, 23:553
Sterling
  silver in, 22:174
Steroids, 22:**851**
  silylation in synthesis of, 22:146
  as veterinary drugs, 24:832
Sterols, 22:852
  in soybeans and other oilseeds, 22:596
Sterox
  ethoxylated alkylphenol surfactant,
    23:511
Stevia
  as sweetener, 23:567
Steviol [471-80-7], 23:568
Stevioside [57817-89-7]
  as sweetener, 23:567
Stickies
  as paper contaminants, 21:11
Stiffeners
  silica sols as, 21:1000
Stigmasterol [83-48-7], 22:853, 863
  in soybeans and other oilseeds, 22:597
Δ^7-Stigmasterol [83-45-4]
  in soybeans and other oilseeds, 22:597
Stilbene [103-30-0], 22:922
Stilbene derivatives, 22:921
Stilbene dyes, 22:**922**
Stilbestrol, 24:833
Stills
  steel for, 22:823
Stimulants, 22:**932**
Stimulated Raman gain spectroscopy
    (srgs), 22:650
Stimulated Raman scattering (srs), 22:650
Stirling engine
  use in solar energy, 22:472
Stirred ball mill, 22:294
Stishovite [13778-37-5], 21:981
Stobbe condensation, 22:1078
Stöber process, 22:79
Stokes' diameters
  of particles, 22:268
Stokes' law, 21:667, 829
  role in particle size measurement,
    22:268
Stokes' lines, 22:649

Stokes' number, 21:674
Stone
  tool materials for, 24:436
Stone treatment
  in textile finishing, 23:905
Stoneware, 22:947
Storage batteries
  use of sulfur for, 23:252
Storage hardening
  of rubber, 21:571
Storax [8046-19-3], 21:299; 22:947
Stormer viscometer, 21:398
Stout, 22:947
Straight, 24:414
Stratospheric ozone protection, 22:532
Strawberries
  mannitol in, 23:97
Strecker reaction, 23:147
Streptococci
  antimicrobial agent for, 24:828
Streptomycin [128-46-1], 22:947; 24:828
Stress cracking, 22:947
Stressing mechanisms
  for size reduction, 22:283
Stresstech rheometer, 21:393
Stretford process
  sulfur recovery, 23:278
Stretford units, 23:444
Strobane [8001-50-1], 22:421
Strontianite, 22:949
Strontium [7440-24-6], 22:947
  complexes with sugar alcohols, 23:103
Strontium-90 [10098-97-2], 22:948
  heat source for thermoelectric devices,
    23:1036
Strontium acetate [543-94-2], 22:952
Strontium–aluminum alloys, 22:950
Strontium and strontium compounds,
    22:**947**
Strontium bromide [10476-81-0], 22:953
Strontium carbonate [1633-05-2], 22:950
Strontium chloride [10476-85-4], 22:948,
    954
Strontium chromate [7789-06-2], 22:953
Strontium fluoride [7783-48-4], 22:954
Strontium hexaferrite [12023-91-5],
    22:953
Strontium hydroxide [18480-07-4], 22:954
Strontium iodide [10476-86-5], 22:954
Strontium nitrate [10042-76-9], 22:950
Strontium oxide [1314-11-0], 22:948
Strontium peroxide [1314-18-7], 22:954

Strontium salicylate [526-26-1], 21:612
Strontium–silicon alloy, 21:1119
Strontium stannate [12143-34-9], 24:128
Strontium sulfide [1314-96-1], 22:952
Strontium titanate [12060-59-2], 22:955; 24:252
Structural foams, 22:955
Structural glazing
    sealant use, 21:665
Structural RIM, 21:198
Structure-activity relationships, 22:955
    for sweeteners, 23:573
Strueverite [12199-39-2]
    tantalum in, 23:660
Strychnine [57-24-9], 22:932
Stupalox, 24:428
Styrene [100-42-5], 22:**956**
    copolymerization with VDC, 24:888
    polymerization in SCFs, 23:470
    polymerization of, 22:1034
    SBR from, 22:996
    sensors for, 21:824
    from toluene, 24:381
    VP copolymerization, 24:1090
Styrene–acrylonitrile (SAN) copolymer, 22:1023
    silane coupling agent for, 22:152
Styrene–butadiene latex, 22:984
Styrene–butadiene rubber (SBR), 21:485; 22:**994**, 984
    reclaiming of, 21:38
    selenium in, 21:713
    sorbitol as stabilizer for, 23:112
    thiols in, 24:28
Styrene–butadiene solution copolymers, 22:1014
Styrene–maleic anhydride copolymers, 22:1048
Styrene plastics, 22:**1015**
4-Styrenesulfonic acid, 23:170
STYREX
    styrene inhibitor, 22:973
Stysanus sp.
    inhibited by sorbates, 22:579
Suberic acid, 22:1073
Succinamic acid [638-32-4], 22:1079
Succinamide [110-14-5], 22:1079
Succinic acid [110-15-6], 22:**1074**
Succinic anhydride [108-30-5], 22:**1074**
Succinimide [123-56-8], 22:1078
Succinyl chloride [543-20-4], 22:1077
Sucralfate [54182-58-0], 23:9

Sucralose [56038-13-2], 23:7, 70, 295
    compared to sucrose, 23:4
    as sweetener, 23:568
Sucrochemicals, 23:5
Sucrochemistry, 23:42, 63
Sucrononic acid [116869-55-7], 23:574
Sucrose [57-50-1], 23:2
    from beets, 23:48
    compared to sucrose, 23:4
    fructose and dextrose in, 23:594
    in soybeans and other oilseeds, 22:598
    substitutes for, 23:556
    from sugarcane, 23:20
    use in density gradient separation, 21:853
Sucrose 6,1',6',-tricarboxylate, 23:73
Sucrose acetals, 23:65
Sucrose derivatives, 23:63
Sucrose 6,6'-dithiol, 23:73
Sucrose 6,6'-episulfide, 23:73
Sucrose monoesters (SMEs), 23:7
Sucrose octaacetate [126-14-7], 23:66
Sucrose octabenzoate [2425-84-5], 23:67
Sucrose polyester (SPE), 23:7
Sugar [57-50-1]
    beet sugar, 23:**44**
    cane sugar, 23:**20**
    centrifugal separation of, 21:868
    properties of sucrose, 23:**1**
    purity, 23:24
    sedimentation in, 21:682
    special sugars, 23:**87**
    substitutes for, 23:556
    sugar analysis, 23:**12**
    sugar derivatives, 23:**63**
    sugar economics, 23:**79**
    use of sulfur dioxide in mfg, 23:310
Sugar alcohols, 23:**93**
    as sweeteners, 23:556
Sugar analysis, 23:12
Sugar balance, worldwide, 23:83
Sugar beets, 23:47
    molasses from, 23:602
    sodium nitrate as fertilizer for, 22:392
Sugarcane, 23:20
    molasses from, 23:602
Sugar consumption, 23:82
Sugar cubes, 23:41
Sugar ethers, 23:64
Sugar production, 23:81
L-Sugars, 23:558
Sugars, special, 23:87

Sugar trading, *23*:13
Sulfa drugs, *23*:119
    veterinary drugs, *24*:828
Sulfamates, *23*:119, 120
Sulfamation, *23*:147, 171
Sulfamic acid [*5329-14-6*], *23*:147, 170,
    195
Sulfamic acid and sulfamates, *23*:**120**
Sulfamoyl chloride [*7778-42-9*], *23*:123
Sulfanes, *23*:284
Sulfanilamide, *23*:133
Sulfanilic acid [*121-57-3*], *23*:133; *24*:15
Sulfated acids, *23*:133
Sulfate process
    for titanium dioxide, *24*:241
Sulfates
    of poly(vinyl alcohol), *24*:989
    in steam, *22*:737
Sulfates, organic, *23*:409
Sulfation, *23*:**146**, 133, 147, 170
    by sulfamic acid, *23*:124
    use of sulfamic acid, *23*:130
Sulfenamides
    vulcanization accelerator, *21*:461
SulFerox process, *23*:444
Sulfides, *23*:134
Sulfinol-M, *23*:440
Sulfinol process, *23*:139, 440
1,1′-Sulfinylbisbenzene [*945-51-7*], *23*:218
1,1′-Sulfinylbisbutane [*2168-93-6*], *23*:218
1,1′-Sulfinylbis(2-chloroethane) [*5819-08-*
    9], *23*:218
1,1′-Sulfinylbisethane [*70-29-1*], *23*:218
2,2′-Sulfinylbisethanol [*3085-45-8*], *23*:217
Sulfinylbismethane [*67-68-5*], *23*:218
1,1′-Sulfinylbis(methylenebenzene)
    [*621-08-9*], *23*:218
1,1′-Sulfinylbispropane [*4253-91-2*], *23*:218
Sulfitation, *23*:147
Sulfitation sugar, *23*:33
Sulfite liquors
    xylose from, *23*:105
Sulfite process, *23*:134
Sulfites, *23*:134
Sulfites, organic, *23*:409
Sulfoacetic acid [*123-43-3*], *24*:2
Sulfoalkylation, *23*:134
Sulfochlorination, *23*:134
    of paraffins, *23*:161
Sulfolane [*126-33-0*], *23*:302, 440
Sulfolane and sulfones, *23*:**134**
Sulfolane-W, *23*:137

Sulfolene [*77-79-2*], *23*:137, 302
α-Sulfomethyl tallowate, *23*:162
Sulfonamide derivative [*2280-49-1*]
    as rubber chemical, *21*:473
Sulfonamides, *23*:146
    as veterinary drugs, *24*:828
Sulfonate AAS
    alkylbenzensulfonate surfactant, *23*:496
Sulfonate esters
    of sugars, *23*:69
Sulfonates, *23*:494
    from sulfonic acids, *23*:198
N-Sulfonation, *23*:147
Sulfonation and sulfation, *23*:**146**
Sulfones, *23*:**134**
Sulfonic acids, *23*:**194**
    naphthalene-based, *23*:206
Sulfonylation
    of toluene, *24*:357
5-Sulfosalicylic acid [*97-05-2*], *21*:605, 617
Sulfotex, *23*:502
    alcohol surfactant, *23*:502
Sulfoxidation
    of paraffins, *23*:161
    of paraffin sulfonates, *23*:172
Sulfoxidation reaction, *23*:302
Sulfoxides, *23*:**217**
Sulfoxylic acid [*20196-46-7*], *23*:312, 319
Sulfreen process, *23*:443
Sulfur [*7704-34-9*], *22*:422; *23*:**232**
    boiling point of, *23*:627
    cross-linking agent, *21*:470
    in pig iron, *22*:767
    recovery from hydrogen sulfide, *23*:434
    refractories for, *21*:79
    removal in steelmaking, *22*:781
    in sodium nitrate, *22*:390
    in steel, *22*:775
    from sulfur dioxide, *23*:446
    sulfuric acid from, *23*:377
    tanks for, *23*:647
Sulfur Black
    thiosulfates as by-product, *24*:58
Sulfur Black 1 [*1326-82-5*], *23*:342, 345,
    348, 352
Sulfur Black 3 [*1326-81-4*], *23*:342
Sulfur Black 9 [*1327-56-6*], *23*:352
Sulfur Black 11 [*1327-14-6*], *23*:342, 353
Sulfur Black T, *23*:342
Sulfur Blue 7 [*1327-57-7*], *23*:352
Sulfur Blue 9 [*1326-97-2*], *23*:342, 352
Sulfur Blue 12 [*1327-96-4*], *23*:342, 353

Sulfur Brown 1 [1326-37-0], 23:342
Sulfur Brown 4 [1326-90-5], 23:351
Sulfur Brown 6 [1327-25-9], 23:351
Sulfur Brown 7 [1327-10-2], 23:351
Sulfur Brown 12 [1327-86-2], 23:350, 355
Sulfur Brown 26 [1326-60-9], 23:350
Sulfur Brown 31 [1327-11-3], 23:351
Sulfur Brown 51 [1327-22-6], 23:342
Sulfur Brown 52 [1327-18-0], 23:342, 350
Sulfur Brown 56 [1327-87-3], 23:351
Sulfur Brown 60 [1327-20-4], 23:342
Sulfur Brown 96 [1326-96-1], 23:348
Sulfurchlorination, 23:430
Sulfur compounds, 23:**267**
    sampling of, 21:635
    for self-assembled monolayers, 23:1095
Sulfur concrete, 23:262
Sulfur dichloride [10545-99-0], 23:269,
    289, 430
Sulfur dioxide [7446-09-5], 23:195, 299,
    340
    emissions control in power generation,
        21:187
    in sulfuric acid, 23:399
    sulfur removal and recovery, 23:446
    use in starch recovery, 22:706
Sulfur dyes, 23:**340**
    thioglycolic acids in, 24:16
    thiosulfates as by-product, 24:58
Sulfur Green 1 [1326-77-8], 23:351
Sulfur Green 2 [1327-74-0], 23:352
Sulfur Green 3 [1327-73-7], 23:342, 352
Sulfur Green 9 [1326-39-2], 23:351
Sulfur Green 11 [12262-52-1], 23:351
Sulfur Green 12 [1236-48-3], 23:350
Sulfur Green 14 [12227-06-4], 23:343
Sulfur hexafluoride [2551-62-4], 23:285
Sulfuric acid [7664-93-9], 23:195, 310, 312,
    363, 366
    boiling point of, 23:627
    sampling in gases, 21:637
    for selenium recovery, 21:694
    in sodium chloride prdn, 22:361
    specific gravity, 23:625
    for sulfonation, 23:147
    from sulfur, 23:258
    from sulfur dioxide, 23:446
Sulfuric acid and sulfur trioxide, 23:**363**
Sulfuric and sulfurous esters, 23:**409**
Sulfurization and sulfurchlorination,
    23:**428**
Sulfur monochloride [10025-67-9], 23:285,
    430

Sulfur nitrides, 23:298
Sulfur Orange 1 [1326-49-4], 23:342, 344,
    350
Sulfur oxides
    sensors for, 21:824
Sulfur polymer cement, 23:262
Sulfur Red 3 [1327-84-0], 23:342, 352
Sulfur Red 6 [1327-85-1], 23:353, 355
Sulfur Red 7 [1327-97-5], 23:342, 353
Sulfur Red 10 [1326-96-1], 23:348, 353
Sulfur removal and recovery, 23:**432**
Sulfur tetrafluoride [7783-60-0], 24:855
Sulfur trioxide [7446-11-9], 23:195, 286,
    363, 364
    for sulfation, 23:147
    for sulfonation, 23:147
Sulfur vulcanization
    of natural rubber, 21:574
Sulfur Yellow 1 [1326-47-2], 23:350
Sulfur Yellow 4 [1326-75-6], 23:341
Sulfur Yellow 9 [1326-40-5], 23:351
Sulfuryl choride [7791-25-5], 23:295
Sulprofox
    thiols in, 24:30
Sultones, 23:197
Sulvanite [15117-74-5], 24:783
Summability equation, 23:1003
Sun
    as power source, 22:465
Sunblock
    titanium dioxide as, 24:239
Suncor process, 23:734
Sunette, 23:563
Sunflowers, 22:591
Sunflower seeds
    extraction using SCFs, 23:467
Sunscreens, 23:452
    salicylates as, 21:615
Sun Yellow [1325-37-7], 22:926
Suosan [140-46-5]
    potential sweetener, 23:573
Supelclean, 24:493
Superabsorbent polymers
    in soil improvement, 22:460
Superacids, 23:210, 452
Superalloys, 23:452
    brazing filler metals for, 22:490
    tool materials for, 24:430
Superclaus process, 23:442
Supercomputers, 23:452
Superconductivity
    of rubidium fullerenes, 21:598

Superconductors
    refractory coatings for, *21*:113
Supercritical fluid chromatography (sfc),
    23:457; *24*:507
Supercritical fluids, *23*:**452**
    for decaffeination of tea, *23*:762
    extraction, *23*:467
    in trace analysis, *24*:495
Supercritical water oxidation, *23*:470
Superfloc, *22*:459
Superfund Amendments and
    Reauthorization Act (SARA), *21*:166;
    24:46
Super III rayon, *24*:162
Superior processing rubbers, *21*:575
Supernovas
    analysis by γ-ray spectroscopy, *22*:656
Superoxides, *23*:477
Superphosphate, *23*:477
Superseed, *22*:950
Super slurper, *22*:460
SUPERTRAP, *23*:456
Supported nickel
    catalysis for sugar alcohols, *23*:105
Suprasil, *21*:1034, 1049, 1058
Suprasil-W, *21*:1034
Surface acoustic wave oscillator
    thin films for, *23*:1087
Surface acoustic waves (SAW)
    sensors based on, *21*:823
Surface analysis
    by infrared spectroscopy, *22*:641
Surface and interface analysis
    using Mössbauer spectroscopy, *22*:656
Surface cleaning
    use of steam, *22*:759
Surface-enhanced Raman scattering
    (sers), *22*:650
Surface lubricants, *21*:207
Surface modification
    organic titanates for, *24*:322
    of self-assembled monolayers, *23*:1092
Surface plasmon resonance, *23*:1087
Surfaces
    of silicates, *22*:10
    silylating agents for interfaces, *22*:151
Surfaces, mineral
    silylating agents for, *22*:147
Surfactants, *23*:**478**, 477; *24*:506
    alkyl hydrogen sulfates, *23*:417
    for cleaning membranes, *21*:317
    effect of supercritical fluids, *23*:462

    fatty acid esters of hexitols, *23*:108
    for monolayer formation, *23*:1077
    in paper recycling, *21*:16
    soap as, *22*:297
    sulfonation and sulfation of, *23*:147
    thiols for, *24*:30
    use in silica prprn, *21*:1019
    use of silicates with, *22*:21
    use of sulfonic acids, *23*:205
Surfactol
    castor oil surfactant, *23*:517
Surfilcon-A, *24*:1079
Surfonic
    alkylphenol surfactant, *23*:503
Surgam [*33005-95-7*], *24*:47
Surgery
    steel for instrument, *22*:822
Surgical gut
    as sutures, *23*:542
Surgidac, *23*:543
Surgigut, *23*:543
Surgilene, *23*:543
Surgilon, *23*:543
Surgipro, *23*:543
Surimi
    sugar alcohols in, *23*:109
Suspensions
    rheological measurements, *21*:427
Sustainable agriculture systems
    reduced pesticide usage, *22*:447
Sutures, *23*:**541**
    silk for, *22*:162
    tantalum for, *23*:671
Sweeteners, *23*:**556**
    compared to sucrose, *23*:4
    corn syrups as, *23*:600
Sweetness
    of selected carbohydrates, *23*:25
    of sugar alcohols, *23*:108
Sweetness potency, *23*:557
Sweet 'n Low, *23*:567
Sweet potato
    starch from, *22*:699
Swirl atomizers, *22*:673
Swiss chard
    sulfoxides in, *23*:217
Switches
    silver for, *22*:176
SYALON, *24*:432
Sylodent 700, *21*:1008
Sylvanite [*1301-81-1*]
    tellurium in, *23*:783

Sylvatic plague, *21*:238
Sympathomimetics, *22*:936
Synchrotrons
  as x-ray sources, *22*:655
Syncrude, *23*:734
Syneresis
  role in sol–gel technology, *22*:508
Syngas, *22*:479
Synroc, *24*:251
Synsil, *21*:1058
Synthetic hydrotalcite [*12304-65-3*]
  as rubber chemical, *21*:473
Synthetic leather
  polyurethanes, *24*:724
Synthetic lubricants, *23*:582
Synthetic organic chemical industry
  (SOCMI), *21*:166
Synthetic rubber, *21*:482, 591; *22*:994
  tall oil resin for, *23*:621
Synthetics
  finishing of, *23*:910
Synthetic surfactants, *22*:321
Synthetic waxes
  as release agents, *21*:210
Syrups, *23*:**582**
  centrifugal separation of, *21*:861
  from sorbitol, *23*:109

T

Table roller mill, *22*:287
Table salt
  iodized, *24*:91
Table sugar, *23*:2
Table wines
  use of sorbates, *22*:582
Tachysterol [*115-61-7*], *22*:857
Tack, *23*:607
Tackifiers
  of adhesives, *24*:971
TackTrap, *21*:255
Taconite, *23*:607
Taguchi gas sensors, *21*:825
Tailings, *23*:734
Takahax process
  sulfur recovery, *23*:278
Talc [*14807-96-6*], *21*:210; *22*:3; *23*:**607**
  as release agent, *21*:210
Talin, *23*:572
Tallic acetate [*2570-63-0*], *23*:959
Tall oil [*8002-26-4*], *23*:**616**

Tall oil fatty acid (TOFA), *23*:616
  surfactants from, *23*:528
Tall oil rosin (TOR), *21*:292; *23*:616
Tallow [*61789-97-7*], *22*:302
Talmage and Fitch
  for sedimentation flux, *21*:673
Talo phosphatation, *23*:35
Tamoxifen [*10540-29-1*], *22*:904
Tanglefoot, *21*:255
Tanks
  refractories for, *21*:84
  sampling in, *21*:638
  sulfamic acid cleaner for, *23*:129
Tanks and pressure vessels, *23*:**623**
Tannase, *23*:761
Tanning agent
  use of titanium compounds, *24*:267
Tannins
  as bird repellents, *21*:254
Tantalic acid [*75397-94-3*], *23*:676
Tantalite [*1306-08-7*]
  magnetic intensity, *21*:876
  tantalum in, *23*:660
Tantalum [*7440-25-7*], *21*:58; *23*:394, 658
  refractory coatings for, *21*:105
  role in steelmaking, *22*:779
  thorium oxides, *24*:76
Tantalum and tantalum compounds,
  23:**658**
Tantalum monocarbide [*12070-06-3*],
  23:676
Tantalum nitride [*12033-62-4*], *23*:676
Tantalum oxide
  as refractory coating, *21*:113
Tantalum(II) oxide [*12035-90-4*], *23*:676
Tantalum pentachloride [*7721-01-9*],
  23:676
Tantalum pentafluoride [*7783-71-3*],
  23:676
Tantalum pentoxide [*1314-61-0*], *23*:665,
  675
Tantalum powders
  for capacitors, *23*:665
Tapeworms
  veterinary drugs against, *24*:832
Tapioca
  dextrose from, *23*:585
  starch from, *22*:699
Tapiolite [*1310-29-8*]
  tantalum in, *23*:660
Tar, *23*:679, 719
  particle-bonding mechanism, *22*:224

specific gravity, 23:625
use in refractory brick, 21:69
Tar and pitch, 23:**679**
Tariffs, 24:527
Tarnish
    silver, 22:169
    silver sulfide, 22:184
Tar sands, 23:**717**
    centrifugal separation of, 21:860
    steam for oil extraction, 22:758
Tartaric acid, 23:745
Tartrazine, 22:190
Taurine [107-35-7], 23:210
Taurocholic acid, 23:210
TBC. See 4-tert-Butylcatechol.
TBS. See 4-tert-Butylstyrene.
TBSI. See N-tert-Butyl-2-benzothiazole-
    sulfenimide.
TBTS. See N-tert-Butyl-2-benzothiazole-
    sulfenamide.
TBzTD. See Tetrabenzylthiuram disulfide.
TCLP. See Toxicity characteristic leaching
    procedure.
TD-5032, 24:148
T-DET
    ethoxylated alkylphenol surfactant,
    23:511
TDF. See Tire-derived fuel.
TDI. See Toluene diisocyanate.
TD resins, 22:116
Tea, 23:**746**
    stimulants in, 22:936
    trace analysis of, 24:515
Tea, instant, 23:753, 761
Teallite [12294-02-9], 24:105, 123
Technetium, 23:768
Technically Specified Rubber (TSR),
    21:565
Technical service, 23:**768**
Technological literacy
    and R&D management, 21:274
Technology management, 21:264
Technology policy
    and R&D management, 21:274
TED. See Transferred electron devices.
TeDEC. See Tellurium diethyldi-
    thiocarbamate.
Teflon, 21:592; 23:389; 24:31
    as sampling probe, 21:634
Telecommunications
    solar energy for, 22:475
Teledium, 23:803

Telephones
    synthetic quartz crystals for, 21:1082
Telephones, cellular
    solid tantalum capacitors in, 23:670
Telescopes
    vitreous silica in prdn of, 21:1068
Television tubes, color
    strontium carbonate in, 22:953
3,5-Telluranedione [24572-07-4], 23:800
Tellurite [14832-87-2]
    tellurium in, 23:783
Tellurium [13494-80-9], 23:782
    impurity in selenium, 21:705
    selenium by-product, 21:696
Tellurium and tellurium compounds,
    23:**782**
Tellurium–copper alloys, 23:802
Tellurium decafluoride [53214-07-6],
    23:796
Tellurium dibromide [7789-54-0], 23:797
Tellurium dichloride [10025-71-5], 23:796
Tellurium diethyldithiocarbamate
    (TeDEC) [20941-65-5], 21:466; 23:788
Tellurium dioxide [7446-07-3], 23:788, 797
Tellurium hexafluoride [7783-80-4],
    23:793
Tellurium nitride [12164-01-0], 23:796
Tellurium oxydibromide [66461-30-1],
    23:797
Tellurium sulfide [16608-21-2], 23:792,
    796
Tellurium sulfite [84074-47-5], 23:786
Tellurium tetrabromide [10031-27-3],
    23:797
Tellurium tetrachloride [10026-07-0],
    23:797
Tellurium tetrafluoride [15192-26-4],
    23:796
Tellurium tetraiodide [7790-48-9], 23:797
Tellurium trioxide [13451-18-8], 23:798
Tellurous acid [10049-23-7], 23:798
Telnic bronze, 23:803
Tempeh, 22:615
Temperature
    determination from optical
    spectroscopy, 22:649
Temperature control
    steam heating for, 22:757
Temperature conversion, 24:637
Temperature measurement, 23:**809**
    thermoelectricity for, 23:1029
Temperature Scale, 23:815

Tempered lead
 sodium in, *22*:347
Tempering
 of steel, *22*:802
Tendons
 silk for, *22*:162
Tenidap [*100599-27-7*], *24*:47
Tenoxicam [*59804-37-4*], *24*:47
Tenting
 textile use, *23*:882
TEOS. See *Tetraethyl orthosilicate.*
Terbium, *23*:832
Terbufos [*13071-79-9*], *22*:422
 thiols in, *24*:30
Terbutylazine, *24*:510
Terephthalic acid, *21*:864; *23*:832
Tergenol
 surfactant, *23*:499
Tergitol, *23*:832
 ethoxylated alcohol surfactant, *23*:509
Terminal sterilization, *22*:846
Terne plate, *23*:832; *24*:117
Terpane
 sampling standards for, *21*:628
Terpenes, *23*:832, 833
C$_{40}$ Terpenes, *23*:874
Terpenoids, *23*:**833**, 832
1,8-Terpin, *23*:835
α-Terpinene [*99-86-5*], *23*:834
γ-Terpinene [*99-85-4*], *23*:834
Terpinene-1-ol [*586-82-3*], *23*:835
Terpinene-4-ol [*562-74-3*], *23*:835
trans-β-Terpineol [*7299-40-3*], *23*:835
α-Terpineol [*98-55-5*], *23*:835
γ-Terpineol [*586-81-2*], *23*:835
Terpinolene [*586-62-9*], *23*:834
α-Terpinyl acetate [*80-26-2*], *23*:855
α-Terpinyl methyl ether [*14576-08-0*],
  23:834
Tertiary amine catalysts
 for flexible foams, *24*:699
Testosterone [*58-22-0*], *22*:859; *24*:834
TETD. See *Tetraethylthiuram disulfide.*
Tetrabenzylthiuram disulfide (TBzTD)
  [*10591-85-2*], *21*:467
Tetrabutoxysilane [*4766-57-8*], *22*:72
Tetra-*n*-butylthiuram disulfide [*1634-02-
  2*], *21*:467
Tetra-*t*-butyl titanate [*3087-39-6*], *24*:275,
  277
Tetracaine hydrochloride [*136-47-0*],
  24:835

Tetracarboxylmethylmercapto-1,4-dithiane
  [*52959-43-0*], *24*:2
Tetracetoxysilane [*5623-90-3*], *22*:72
2,3,7,8-Tetrachlorodibenzodioxin [*1746-
  01-6*], *24*:869
2,3,7,8-Tetrachlorodibenzofuran [*51207-
  31-9*], *24*:869
1,1,2,2-Tetrachloroethane [*79-34-5*]
 as HAP compound, *22*:532
 as solvent, *22*:538
Tetrachloroethane [*79-34-5*]
 vinylidene chloride from, *24*:884
Tetrachloroethylene [*127-18-4*], *24*:877
 regulatory level, *21*:160
Tetrachlorosilane [*10026-04-7*], *22*:31, 61
 role in sol–gel technology, *22*:515
Tetracresyl titanate [*28503-70-0*], *24*:276
Tetracyclines, *23*:882
 trace and residue analysis of, *24*:503
 as veterinary drugs, *24*:829
Tetradecamethylhexasiloxane [*107-52-8*],
  22:103
n-Tetradecyl hydrogen sulfate [*4754-44-3*],
  23:410
Tetradimethylamino titanate [*3275-24-9*],
  24:277
Tetradymite [*1304-78-5*]
 tellurium in, *23*:783
Tetra(3-eicosylpyridinium)porphyrin
  bromide
 thin films of, *23*:1083
Tetraethoxysilane [*78-10-4*], *22*:72
Tetraethyllead
 from sodium, *22*:344
Tetraethyl orthosilicate (TEOS) [*78-10-4*],
  22:69, 498
 vitreous silica from, *21*:1037
Tetraethylthiuram disulfide (TETD)
  [*97-77-8*], *21*:467
Tetrafilcon-A, *24*:1079
Tetrafluoroethylene
 polymerization in SCFs, *23*:470
Tetrafluorosilane [*7783-61-1*], *22*:33, 61
Tetrahedrite [*12054-35-2*], *22*:170
Tetra-*n*-hexadecyl titanate [*34729-16-3*],
  24:285
Tetrahexafluoroisopropyl titanate
  [*21416-30-8*], *24*:277
Tetra-*n*-hexyl titanate [*7360-52-3*], *24*:285
Tetrahydrofuran [*109-99-9*]
 polymerization of, *23*:209
 as solvent, *22*:548
 from succinic anhydride, *22*:1077

Tetrahydrofurfuryl alcohol [97-99-4]
  as solvent, 22:536
Tetrahydrolinalool [78-69-3], 23:850, 856
Tetrahydromyrcenol [18479-57-7], 23:850
Tetrahydrotellurophene [3465-99-4],
  23:800
Tetrahydrothiophene [110-01-0], 24:40
Tetrahydrothiophene dioxide, 23:440
Tetrahydrothiophene-1,1-dioxide, 23:134
Tetraiodosilane [13465-84-4], 22:33
Tetraisobutylthiuram disulfide [3064-73-
  1], 21:467
Tetraisobutylthiuram monosulfide
  as rubber chemical, 21:473
Tetraisopropoxysilane [1992-48-9], 22:72
Tetrakisazo dyes
  stilbene dye examples, 22:927
Tetrakis(cumylphenoxy)phthalocyanine
  thin films of, 23:1083
Tetrakis(2-ethylbutoxy)silane [78-13-7],
  22:72
Tetrakis(2-ethylhexoxy)silane [115-82-2],
  22:72
Tetrakis[hydroxyisopropyl]ethylene-
  diamine, 24:296
Tetrakis(2-methoxyethoxy)silane [2157-
  45-1], 22:72
Tetrakis(s-butoxy)silane [5089-76-9], 22:72
Tetramethoxysilane [681-84-5], 22:72
Tetramethyldisiloxane, 23:1066
Tetramethylene sulfone, 23:134
Tetramethyllead
  from sodium, 22:344
Tetramethylthiuram disulfide [137-26-8],
  21:467
  as deer repellent, 21:257
Tetramethylthiuram monosulfide
  [97-74-5], 21:467
Tetramethyl titanate [992-92-7], 24:278
m-Tetramethylxylylene [58067-42-8],
  24:707
Tetramisole [6649-23-6], 24:830
Tetraneopentyltitanium [36945-13-8],
  24:309
Tetranorbornyltitanium [36333-76-3],
  24:309
Tetra[4-oxy(2-docosanoic acid)]phenyl-
  porphyrin
  thin films of, 23:1083
Tetra-n-pentyl titanate [10585-24-7],
  24:285
Tetraphenoxysilane [1174-72-7], 22:72

Tetraphenylporphyrin palladium
  LB films of, 23:1087
Tetraphenylsilane [1048-08-4], 22:60
Tetraphenyl telluride [64109-07-5], 23:800
Tetraphenyltin [595-90-4], 24:132
Tetrapropoxysilane [682-01-9], 22:72
Tetraterpenes, 23:833, 874
Tetrazotized 4,4′-diamino-2,2′-stilbenedi-
  sulfonic acid [57153-16-9], 22:925
Tetronic polyols, 23:528
Tevdek II, 23:543
TEWI. See Total equivalent warming
  impact.
Textile applications
  sodium dithionite use, 23:319
Textile finishing, 23:890
Textile mechanics, 24:171
Textile processing
  sodium sulfite used in, 23:314
Textile reinforcements
  for tire cords, 24:161
Textiles, 23:**882**
  finishing, 23:**890**
  PVC use, 24:1040
  reverse osmosis for wastewater, 21:322
  silk for, 22:155
  steel for manufacture of, 22:823
  sulfolane in fabrication, 23:140
  sulfonic acid-derived dyes in, 23:206
  sulfur dyes for, 23:360
  sulfur in prdn of, 23:259
  testing, 23:**916**
  use of hexitols in, 23:107
Textile sizing
  PVA resins as, 24:980
Tex-Wet
  dialkyl sulfosuccinate surfactant,
  23:498
TFT. See Thin-film transistor.
TGS. See Triglycine sulfate.
Thallic bromide [13701-90-1], 23:955
Thallic chloride [13453-32-2], 23:955
Thallic fluoride [7783-57-5], 23:955
Thallic nitrate [13746-98-0], 23:955
Thallic oxide [1314-32-5], 23:955
Thallium [7440-28-0], 23:952
  in shape-memory alloys, 21:964
Thallium and thallium compounds, 23:**952**
Thallium(III) bromide tetrahydrate
  [13453-29-7], 23:956
Thallium(I) chloride [7791-12-0], 23:952
Thallium(I) ethoxide [20398-06-5], 23:956

Thallium(I) hydroxide [12026-06-1], 23:952

Thallium(I) oxide [1314-12-1], 23:952

Thallium(III) oxide [1314-32-5], 23:953

Thallium(I) sulfate [7446-18-6], 23:952

Thallium(I) tetrahydridoborate [61204-71-5], 23:956

Thallium(III) trifluoroacetate [23586-53-1], 23:958

Thallium(I) triiodide [13453-37-7], 23:956

Thallotoxicosis, 23:961

Thallous acetate [563-68-8], 23:955

Thallous bicarbonate [65975-01-1], 23:954

Thallous bromide [7789-40-4], 23:955

Thallous carbonate [29809-42-5], 23:955

Thallous chloride [7791-12-0], 23:955

Thallous ethoxide [20398-06-5], 23:957

Thallous fluoride [7789-27-7], 23:955

Thallous formate [992-98-3], 23:955

Thallous hydroxide [12026-06-1], 23:955

Thallous iodide [7790-30-9], 23:955

Thallous nitrate [10102-45-1], 23:955

Thallous oxide [1314-12-1], 23:955

Thallous sulfate [7446-18-6], 23:955

Thaumatin [53850-34-3]
    as sweetener, 23:572

Theaflavin, 23:755

Theanine [3081-61-6]
    in tea, 23:750

Thearubigens, 23:756

Theasinensin A, 23:756

Theasinensin F, 23:756

Thenardite, 22:403

Theobromine [83-67-0], 23:750

Theophylline [58-55-9], 23:750

Thermal electric technology
    from solar energy, 22:471

Thermal pollution, 23:**963**

Thermistors
    role in temperature measurement, 23:819

Thermocouples, 21:1065; 23:821
    for optical spectroscopy, 22:635
    refractory coatings for, 21:111
    rhenium and rhenium alloys in, 21:341
    for thermoelectric energy conversion, 23:1034

Thermodynamics, 23:**985**

Thermoelectric energy conversion, 23:**1027**

Thermoelectrics
    tellurium in, 23:791

Thermographic dyes, 23:1040

Thermometers, 23:811

Thermopave, 23:261

Thermoplastic elastomer, 24:1033

Thermoplastic natural rubber, 21:577

Thermoplastic polymers
    organic tire cords, 24:168
    for sutures, 23:550

Thermoplastic polyolefin (TPO)
    as roofing material, 21:449

Thermoplastic polyurethane elastomers (TPUs), 24:696

Thermoplastics
    as roofing material, 21:448

Thermos bottles, 22:176

Thermoset flexible polyurethane, 24:702

Thermoset polyurethanes, 24:717

Thermoset processes
    for reinforced plastics, 21:196

Thermowells
    tantalum in, 23:671

Thiabendazole [148-79-8], 24:831

2-Thiapropane, 24:23

Thiazole dyes, 23:1040

Thickeners, 23:171
    starches as, 22:714

Thickening, 21:667

Thielavia basicola
    inhibited by sorbates, 22:579

Thieno[3,4-d]isothiazolone dioxide [59337-79-0]
    sweetener, 23:566

2-Thienyllithium [2786-07-4], 24:39

Thin-film method
    for sulfation, 23:171

Thin films
    analysis using x-ray fluorescence, 22:655
    ceramic from sol–gel technology, 22:523
    film formation techniques, 23:**1040**
    monomolecular layers, 23:**1077**
    of organic titanates, 24:322
    of stannic oxide, 24:126
    on tantalum, 23:673
    thorium phosphates as, 24:77
    of vitreous silica, 21:1000, 1067

Thin-film transistor (TFT), 21:757

Thin-layer chromatography
    of sugar alcohols, 23:103
    for trace analysis, 24:498

Thiobacillus ferroidans
    use in sulfur recovery, 23:445

4,4'-Thiobisphenol [2664-63-3]
    insecticide intermediate, 23:291
5,5'-Thiobis-1,3,4-thiadiazole-2(3H)thione
    [7340-97-8]
    cross-linking agent, 21:470
Thiocarbanilide, 21:467
Thiocyanates, 23:1118
Thiocyanic acid [463-56-9], 23:319
Thiocyanomethylthiobenzothiazole
    [21564-17-0], 23:323
Thiodiglycol, 24:24
Thiodiglycolic acid [123-93-3], 24:3, 7
Thiodipropionic acid [111-17-1], 24:7, 16
Thioglycolic acid [68-11-1], 24:1, 8
Thiokol, 21:491
Thiolactic acid, 24:6
Thiols, 24:19
    silver complexes of, 22:186
Thiomaleic anhydride [6007-87-0], 24:40
Thiomalic acid [70-49-5], 24:7
Thionol Black BM
    sulfur dye, 23:343
Thionol Brilliant Green GG
    sulfur dye, 23:342
Thionol Ultra Green B
    sulfur dye, 23:343
Thionyl bromide [507-16-4], 23:292
Thionyl chloride [7719-09-7], 23:197, 291
    from sulfur monochloride, 23:287
Thiopaq Bioscrubber process, 23:449
Thiopental [77-27-0], 24:835
2-Thiopheneacetic acid [1918-77-0], 24:40,
    44
2-Thiopheneacetonitrile [20893-30-5],
    24:40
Thiophene and thiophene derivatives,
    24:34, 36, 37, 854
2-Thiophenecarboxaldehyde [98-03-3],
    24:39, 43, 45, 48
3-Thiophenecarboxaldehyde [498-62-4],
    24:43
2-Thiophenecarboxylic acid [527-72-0],
    24:39
2-Thiopheneglyoxylic acid [4075-59-6],
    24:44
3-Thiophenemalonic acid [21080-92-2],
    24:48
Thiophene-2-thiol [7774-74-5], 24:39
Thiophosgene [463-71-8], 23:271
Thiosalicylic acid [147-93-3], 21:623
Thiosulfates, 24:51
    silver complexes of, 22:185

Thiosulfuric acid [13686-28-7], 23:312
Thiourea [62-56-6], 23:280
Thioureas
    silver complexes of, 22:185
    vulcanization accelerator, 21:463
Thiouricil
    antithyroid prprn, 24:99
Thiuram [137-26-8], 22:400
    vulcanization accelerator, 21:462
Thixotropic materials, 21:353
    silica gels as, 21:1023
Thixotropy, 21:351
Thomson effect, 23:1030
Thoria
    in refractories, 21:78
    in rhenium alloys, 21:340
Thorianite, 24:70
Thorite, 24:70
Thorium [7440-29-1], 24:68
Thorium and thorium compounds, 24:68
Thorium chloride [10026-08-1], 24:72
Thorium fluoride [13709-56-6], 24:73
Thorium hydroxide [13825-36-0], 24:76
Thorium iodide [7790-49-0], 24:72
Thorium oxide [1314-20-1], 21:58; 24:72
    as refractory material, 21:57
Thorium perchlorate [16045-17-3], 24:73
Thorium(IV) phosphates [15578-50-4],
    24:77
Thorocene [12702-09-9], 24:83
Thorotrast, 24:85
Thortveitite [17442-06-7], 22:3
Thread guides
    titanium dioxide in, 24:238
Threitol [7493-90-5], 23:95
D-Threitol [2418-52-2], 23:95
D,L-Threitol [6968-16-7], 23:95
L-Threitol [2319-57-5], 23:95
Threonine
    in silk, 22:156
    in soybeans and other oilseeds, 22:596
Threshold planning quantity (TPQ),
    21:166
α-Thrombin
    grown in space, 22:622
Thrombolytic agents, 24:88
Thulium, 24:88
Thymol [89-83-8], 23:859
Thymolphthalein, 22:418
Thyrocalcitonin, 24:89
Thyroglobulin [9010-34-8], 24:94
Thyroid and antithyroid preparations,
    24:89

Thyroid hormone-binding globulin [9010-34-8], 24:90

Thyroid hormone-binding prealbumin [632-79-1], 24:90

Thyroid-stimulating hormone [9002-71-5], 24:89

Thyroliberin, 24:89

Thyrotoxicosis, 24:91

Thyrotropin, 24:89

D-Thyroxine [51-49-0], 24:89

L-Thyroxine [51-48-9], 24:89

Ticarcillin [34787-01-4], 24:48

Ticlopidine [55142-85-3], 24:47

TI.Cron, 23:543

Tidal power, 21:235

Tigogenin [77-60-1], 22:862

Tillage practices
  and pesticide runoff, 22:443

Tilmecosen
  as veterinary drugs, 24:827

Timepidium Bromide [35035-05-3], 24:48

Time–temperature–transformation (TTT) diagram
  for steel, 22:795

Time-weighted average, 21:167

Timing control
  synthetic quartz crystals for, 21:1082

Tin [7440-31-5], 23:1045; 24:105
  in brazing filler metals, 22:492
  plating with methanesulfonic acid, 23:326
  in scrap for steel, 22:779
  tantalum as byproduct of, 23:660
  vapor pressure of, 23:1046

Tin alloys, 24:105
  as shape-memory alloys, 21:964
  as solders, 22:485

Tin compounds, 24:122

Tin crystals, 24:124

Tinctures of iodine, 22:847

Tin–lead alloys, 24:129
  as solders, 22:487

Tin oxide
  as refractory coating, 21:113
  in sensors, 21:825
  thin films of, 23:1073

Tin plating, 24:115, 129

Tin slags
  tantalum from, 23:661

Tin–sodium alloys, 22:347

Tin Sol process, 24:128

Tinstone, 24:122

Tinted sealers, 22:695

Tin tetrachloride, 22:54

Tinti and Nofre sweetener model, 23:574

Tinuvin P
  PS stabilizer, 22:1023

Tin–zinc alloys
  as coatings, 24:117

TiO, 24:428

Tiocarbazil
  thiols in, 24:30

Tioconazole [65899-73-2], 24:48

Tipepidine [5169-78-8], 24:48

Tiquizium Bromide [71731-58-3], 24:48

Tire cord, 24:161

Tire-derived fuel (TDF), 21:24, 27, 231

Tire reinforcement
  textile use, 23:882

Tire-reinforcement fibers, 24:163

Tires
  compounding, 21:520
  natural rubber in, 21:580
  precipitated silica in, 21:1025
  recycling, 21:23
  SBR use, 22:1011
  silica for, 21:1001

Tirilazad mesylate [110101-67-2], 22:908

Tissue repairs
  silk for, 22:162

Titania
  dopant for vitreous silica, 21:1052
  in sol–gel polymer hybrids, 22:526

Titanic amide [15792-80-0], 24:232

Titaniferrous magnetite
  magnetic intensity, 21:876

Titanite, 24:264

Titanium [7440-32-6], 23:1045; 24:186
  in brazing filler metals, 22:492
  brazing of, 22:484
  for centrifuges, 21:845
  in ferrous shape-memory alloys, 21:965
  in silicon and silica alloys, 21:1106
  in steels, 22:749, 779, 811
  thorium oxides, 24:76
  tin alloys of, 24:120
  vapor pressure of, 23:1046
  vitreous silica impurity, 21:1036

Titanium aluminides, 22:824
  as refractory coatings, 21:99

Titanium and titanium alloys, 24:186
  as shape-memory alloys, 21:964
  vanadium in, 24:794

Titanium boride
  in tool materials, 24:430

Titanium bronzes, *24*:251

Titanium carbide [*12070-08-05*], *24*:228
  as refractory coatings, *21*:95
  in steel tool materials, *24*:403
  thin films of, *23*:1061

Titanium carbide powder, *24*:230

Titanium carbonitride, *24*:415

Titanium compounds, *24*:**225**
  inorganic, *24*:**225**
  organic, *24*:**275**

Titanium diboride [*12405-65-35*], *24*:227
  refractory coating, *21*:113
  thin films of, *23*:1061

Titanium dibromide [*13873-04-5*], *24*:261

Titanium dichloride [*10049-06-6*], *24*:256

Titanium difluoride [*13814-20-5*], *24*:254

Titanium diiodide, *24*:262

Titanium dioxide [*13463-67-7*], *24*:235
  PVC pigment, *24*:1035
  sulfur in prdn of, *23*:259
  thin films of, *23*:1053
  use of sulfuric acid, *23*:397
  vitreous silica dopant, *21*:1068

Titanium disilicide [*12039-83-7*], *24*:263

Titanium disulfide [*12039-13-3*], *24*:230,
  265

Titanium hydrides, *24*:227

Titanium isopropoxide [*546-89-9*], *22*:57;
  24:238

Titanium monophosphide [*12037-65-9*],
  24:264

Titanium monosulfide [*12039-07-5*],
  24:265

Titanium monoxide [*12137-20-1*], *24*:233

Titanium monoxychloride, *24*:261

Titanium nitrate [*13860-02-1*], *24*:232

Titanium nitride [*25583-20-4*], *24*:231
  coating for tools, *24*:403, 418
  as refractory coating, *21*:95, 114
  thin films of, *23*:1052

Titanium oxide [*13463-67-7*], *21*:58
  as refractory material, *21*:57

Titanium oxide dichloride [*13780-39-8*],
  24:261

Titanium(III) phosphate [*24704-65-2*],
  24:264

Titanium(IV) phosphate gel [*17017-60-6*],
  24:265

Titanium 1,3-propylenedioxide bis(ethyl
  acetoacetate) [*36497-11-7*], *24*:293

Titanium pyrophosphate [*13470-09-2*],
  24:265

Titanium sesquioxide, *24*:233

Titanium sesquisulfide [*12039-16-6*],
  24:265

Titanium silicates, *24*:303

Titanium–silicon alloy, *21*:1119

Titanium subsulfide [*1203-08-6*], *24*:265

Titanium(IV) sulfate, *24*:267

Titanium(II) sulfide [*12039-07-5*]
  in steel, *22*:816

Titanium tetrabromide [*7789-68-6*], *24*:262

Titanium tetrachloride [*7550-45-0*],
  24:257
  for chemical vapor deposition, *23*:1060
  organic titanium compounds from,
    24:275

Titanium tetrafluoride [*7783-63-3*], *24*:255

Titanium tetraiodide [*7720-83-4*], *24*:262

Titanium tribromide [*13135-31-4*], *24*:261

Titanium trichloride [*7705-07-9*], *24*:257

Titanium trichloride hexahydrate
  [*19114-57-9*], *24*:257

Titanium trifluoride [*13470-08-1*], *24*:255

Titanium triiodide, *24*:262

Titanium triphosphide [*12037-66-0*],
  24:264

Titanium trisulfide [*12423-80-2*], *24*:266

Titanium tubing
  brazing filler metals for, *22*:490

Titanium–zirconium-base alloys
  as brazing filler metals, *22*:495

Titanocene dichloride, *24*:307

Titanous amide [*15190-25-9*], *24*:232

Titanous sulfate [*10343-61-0*], *24*:266

Titanoxanes, *24*:281

Titanyl sulfate [*13825-74-6*], *24*:267

Titratable base number
  of a lube sulfonate, *23*:165

TMS triflate [*27607-77-8*], *22*:145

Tobacco
  molasses in curing of, *23*:604
  sodium nitrate as fertilizer for, *22*:392

Tobias acid, *24*:350

Tocopherols, *24*:350

TOFA. See *Tall oil fatty acid.*

Toffees
  molasses in, *23*:604

Tofu, *22*:615

Toilet bowls
  sulfamic acid cleaner for, *23*:129

Toilet preparations, *24*:350

Toiletry products
  saccharin in, *23*:566

Tolnaftate, *23*:272

Tolu balsam, *24*:350

Toluene [*108-88-3*], *24*:**350**, 856
  as HAP compound, *22*:532
  as solvent, *22*:540
  sulfonation of, *23*:159
  supercritical fluid, *23*:454

Toluenediamines, *24*:389

Toluene diisocyanate (TDI) [*1321-38-6*],
    24:696, 706
  in grouts, *22*:456
  in urethane sealant, *21*:657

p-Toluenesulfonic acid [*104-15-4*], *23*:194;
    24:386

p-Toluenesulfonyl chlorides, *24*:386

α-Toluenethiol [*100-53-8*], *24*:21

Toluidines [*26915-12-8*]
  as solvent, *22*:540

Tolylfluanid [*731-27-1*], *23*:275

Tomatidine [*77-59-8*], *22*:863

Tomatoes
  sulfoxides in, *23*:217

Tone, *22*:323

Toners, *22*:695

Tool materials, *24*:**390**
  titanium carbide in, *24*:231

Tools
  refractory coatings for, *21*:113
  steel for, *22*:817

Tools, cutting
  tantalum carbide in, *23*:676

Tooth decay
  prevention using sugar alcohols, *23*:107

Toothpaste, *24*:456
  precipitated silica in, *21*:1025
  silica in, *21*:1001
  sorbitol in, *23*:111

TOR. See *Tall oil rosin.*

Torbernite [*26283-21-6*], *24*:643

Torpex, *24*:456

Torque, *24*:139

Torsional Braid Analyzer, *21*:416

Torsional pendulum, *21*:412

Torulaspora rosei
  inhibited by sorbates, *22*:580

Torulopsis candida
  inhibited by sorbates, *22*:580

Torulopsis caroliniana
  inhibited by sorbates, *22*:580

Torulopsis minor
  inhibited by sorbates, *22*:580

Torulopsis polcherrima
  inhibited by sorbates, *22*:580

Torulopsis versitalis lipofera
  inhibited by sorbates, *22*:580

Total Equivalent Warming Impact
    (TEWI), *21*:133

Total organic carbon
  separation by reverse osmosis, *21*:323

Total suspended particulates (TSP),
    21:167

Total suspended solids (TSS), *21*:167

Tourmaline
  magnetic intensity, *21*:876

Tows
  in reinforced plastics, *21*:195

Toxaphene [*8001-35-2*], *22*:421; *24*:456
  regulatory level, *21*:160

Toxic effects, *24*:458
  biochemical uncoupling, *24*:463
  degeneration, *24*:461
  endocrine system disruption, *24*:463
  enzyme inhibition, *24*:463
  hypersensitivity, *24*:461
  immunosuppression, *24*:462
  inflammation, *24*:461
  mutagenesis, *24*:462
  necrosis, *24*:461
  neoplasia, *24*:462
  sensory irritation, *24*:463
  teratogenesis, *24*:463

Toxicity
  fate of absorbed chemicals, *24*:467
  hazard evaluation, *24*:485
  testing procedures, *24*:477

Toxicity characteristic leaching procedure
    (TCLP), *21*:166

Toxicity, oral
  for sugar alcohols, *23*:106

Toxicity studies, *24*:464

Toxicology, *24*:**456**

Toxic Release Inventory (TRI), *21*:167

Toxic Substances Control Act, *21*:167

Toxoid, *24*:728

Toxoiding, *24*:731

Toyocat, *24*:700

TPO. See *Thermoplastic polyolefin.*

TPQ. See *Threshold planning quantity.*

TPUs. See *Thermoplastic polyurethane
    elastomers.*

Trace and residue analysis, *24*:**491**
  using x-ray fluorescence, *22*:655

Trace metals
  use of atomic fluorescence spectroscopy,
    22:653

Tracers, 24:523
  thorium isotopes for, 24:70
Tracheography
  use of tantalum in, 23:671
Tragacanth, 24:523
Training
  technical service, 23:774
Tramp iron
  magnetic separators, 21:878
Tranquilizers, 24:523
  in veterinary medicine, 24:834
Transesterification, 23:101
  titanate catalysis, 24:325
Transferred electron devices (TED),
    21:776
Transformer oils
  purification by centrifuge, 21:857
Transformers
  steel for, 22:826
Transistors, 21:732; 24:523
  pure silicon for, 21:1102
Transistor technology, 21:720
Transit 5BN-1, 23:1035
Transit 5BN-2, 23:1036
Transite
  for cooling tower packing, 22:213
Transmission electron microscopy
  for sol–gel technology studies, 22:507
Transportation, 24:**524**
  steel equipment for, 22:823
Transport phenomena
  in MOCVD, 21:771
Transuranium elements, 24:550
Tranylcypromine [155-09-9], 22:939
Trazodone [19794-93-5], 22:945
Treacle, 23:25, 42
Tread rubber
  SBR use, 22:1011
Treadwear, 24:178
Treatment, storage, and disposal facility
    (TSDF), 21:167
Tremolite [14567-73-8], 22:3
Trenbolone [10161-33-8], 24:834
TRI. See *Toxic Release Inventory*.
Triad, 23:1036
Triamcinolone [124-94-7], 22:884; 24:832
Triarylmethane dyes, 24:551
Triazinetriol, 24:550
1,2,3-Triazoles
  silylating agents in synthesis of, 22:146
Triazones
  in textile finishing, 23:899

Tribenzyltin chloride [3151-41-5], 24:138
2,4,6-Tribromophenol [25376-38-9], 21:606
Tribromosilane [7789-57-3]
  for microelectronics, 22:47
Tributyl phosphate
  for tantalum extraction, 23:663
  for thorium recovery, 24:70
Tri-n-butyl phosphate [126-73-6]
  added to supercritical CO_2, 23:468
Tri-t-butylsilane [18159-55-2], 22:59
Tributyltin acrylate [13331-52-7], 24:140
Tricalcium phosphate
  salt additive, 22:368
Trichlorobenzene
  rejection by reverse osmosis membrane,
    21:319
1,2,4-Trichlorobenzene [120-82-1]
  as solvent, 22:538
1,1,3-Trichlorobutane [13279-87-3], 24:855
1,1,1-Trichloroethane [71-55-6], 24:867
  as HAP compound, 22:532
  as solvent, 22:538
1,1,2-Trichloroethane [79-00-5], 24:855
  as HAP compound, 22:532
  as solvent, 22:538
  vinylidene chloride from, 24:884
Trichloroethyl acetate [625-24-1]
  vinylidene chloride from, 24:884
Trichloroethylene [79-01-6], 24:865
  as HAP compound, 22:532
  regulatory level, 21:160
  as solvent, 22:538
Trichloromethanesulfenyl chloride
    [594-42-3], 23:272
Trichlorophenol
  rejection by reverse osmosis membrane,
    21:319
2,4,5-Trichlorophenol [95-95-4]
  regulatory level, 21:160
2,4,6-Trichlorophenol [88-06-2]
  regulatory level, 21:160
  rejection by reverse osmosis membrane,
    21:319
2,4,5-Trichlorophenoxy [93-72-1]
  regulatory level, 21:160
Trichlorosilane [10025-78-2], 21:1087;
    22:31, 38; 24:854
  use in silicon purification, 21:1092
3,3,4-Trichlorosulfolane [42829-14-1],
    23:135
4,1',6'-Trichloro-4,1',6'-trideoxygalacto-
    sucrose [56038-13-2], 23:70

1,1,2-Trichloro-1,2,2-trifluoroethane [76-13-1]
  as solvent, 22:538
Trichoderma viride
  inhibited by sorbates, 22:579
ω-Tricosenoic acid
  thin films of, 23:1081
Tricresyl phosphate, 24:550
Tricyclene [508-32-7], 23:834, 839
Tricyclohexylsilane [1629-47-6], 22:53
Tridymite [15468-32-3], 21:981, 1077
  in silica refractories, 21:83
Triethanolamine [102-71-6], 24:295, 550
  as grouting catalyst, 22:455
  as solvent, 22:540
Triethoxysilane [998-30-1], 22:48, 51, 70
Triethoxytitanium [19726-75-1], 24:304
Triethylamine [121-44-8]
  as solvent, 22:540
Triethylbromosilane [1112-48-7], 22:54
Triethyldisilazane [2117-18-2], 22:52
Triethylene glycol [112-27-6]
  as solvent, 22:546
Triethylene glycol dinitrate, 24:550
Triethylene glycol divinyl ether [76-12-8], 24:1066
Triethylsilane [617-86-7], 22:38
Triflic acid [1493-13-6], 23:209
Trifluoroacetic acid, 22:56
Trifluoromethane (ethyl) thioglycolate, 24:16
Trifluoromethanesulfonic acid [1493-13-6], 23:194, 195
(3,3,3-Trifluoropropyl)trimethoxysilane
  critical surface tension of, 22:148
Trifluorosilane [13465-71-9], 22:43
Trifluralin [1582-09-8], 22:422, 427
Triglycerides
  extraction using SCFs, 23:467
  in soybeans and other oilseeds, 22:596
  titanate catalysis, 24:325
Triglycine sulfate (TGS) crystals
  grown in space, 22:622
L-Triiodothyronine [6893-02-3], 24:89
Triisopropanolamine [122-20-3], 24:295
Triisopropylchlorosilane [13154-24-0], 22:54
Trimene base, 24:550
Trimethoprim [738-70-5], 24:828
Trimethoprim sulfa
  as veterinary drugs, 24:827
Trimethoxysilane [2487-90-3], 22:48, 70

3-(Trimethoxysilyl)propyl methacrylate, 22:525
endo-1,7,7-Trimethylbicyclo[2.2.1]heptan-2-ol [507-70-0], 23:853
exo-1,7,7-Trimethylbicyclo[2.2.1]heptan-2-ol [124-76-5], 23:846
Trimethylchlorosilane [75-77-4]
  silylating agent, 22:144
α,3,3-Trimethylcyclohexane methanol formate [25225-08-5], 23:851
Trimethylolethane, 24:550
Trimethylolethane trinitrate., 24:550
Trimethylolpropane, 24:550
2,2,4-Trimethylpentane [540-84-1]
  as HAP compound, 22:532
2,2,4-Trimethyl-1,3-pentanediol, 24:290
2,4,4-Trimethyl-2-pentanethiol [141-59-3], 24:21
N-Trimethylsilylacetamide [13435-12-6]
  silylating agent, 22:144
Trimethylsilylazide [4648-54-8], 22:146
N-Trimethylsilyldiethylamine [996-50-9]
  silylating agent, 22:144
N-Trimethylsilylimidazole [18156-74-6]
  silylating agent, 22:144
Trimethylsilyl iodide [16029-98-4]
  silylating agent, 22:144
Trimethylsilyl trifluoromethanesulfonate [27607-77-8]
  silylating agent, 22:144
Trimethylsulfoxonium iodide, 23:223
Trimethyltellurium iodide [18987-26-3], 23:800
Trimethylthiourea [2489-77-2], 21:468
Trimipramine [739-71-9], 22:941
2,4,6-Trinitrophenol [88-89-1]
  from salicylic acid, 21:605
Tri-n-octylphosphine oxide
  for tantalum extraction, 23:663
Trioxane, 24:550
Tripentaerythritol, 24:550
N,N',N''-Triphenylaminotriphenyl-methane hydrochloride [2152-64-9], 24:556
Triphenyldisilazane [4158-64-9], 22:52
Triphenylmethane and related dyes, 24:**551**
Triphenylphosphine [603-35-0], 23:196
Triphenylsilicide [15487-82-8], 22:55
Triphenyltin acetate [900-95-8], 24:138
Triphenyltin hydroxide [76-87-9], 24:138
Tripoli, 21:1028

Tripropylene glycol [24800-44-0]
   as solvent, 22:546
Tris
   in textile finishing, 23:908
2,4,6-Tris(N,N-dimethylaminomethyl)-
      phenol (DMT-30), 24:698
1,3,5-Tris(3-dimethylaminopropyl)-
      hexahydro-s-triazine, 24:698
Trisilane [7783-26-8], 24:854
Trisilylamine [13862-16-3]
   for microelectronics, 22:47
Trisulfuric acid, 23:373
Triterpene, 23:833
Trithiocyanuric acid [638-16-4]
   cross-linking agent, 21:470
Trititanium pentoxide, 24:234
Tritium [15086-10-9]
   as components of steam, 22:721
Triton RW, 23:527
   ethoxylate surfactant, 23:526
Triton X
   alkylphenol surfactant, 23:503
6,1',6'-Tri-O-tritylsucrose [35674-14-7],
      23:64
Triuranium octaoxide [1344-59-8], 24:663
Trolox, 23:763
Trosofoil, 24:932
Troubleshooting
   technical service, 23:772
Truncatella sp.
   inhibited by sorbates, 22:579
Trycol
   ethoxylated alcohol surfactant, 23:509
Trylox
   castor oil surfactant, 23:517
Trymeen
   ethoxylate surfactant, 23:526
Tryosine
   in soybeans and other oilseeds, 22:596
Trypanosomiasis
   developing vaccines for, 24:735
Trypsin, 24:572
Trypsin inhibitors
   in soybeans and other oilseeds, 22:596
Tryptophan, 24:572
   in silk, 22:156
   in soybeans and other oilseeds, 22:596
TSDF. See Treatment, storage, and
      disposal facility.
TSP. See Total suspended particulates.
TSR. See Technically Specified Rubber.
TSS. See Total suspended solids.

T-2 toxin, 22:600
TTT diagram. See Time–temperature–
      transformation diagram.
Tuads, 24:572
Tubers
   starch from, 22:699
Tuliokite [128706-42-3], 24:77
Tumbling, 22:229
Tung oil, 24:572
Tungsten [7440-33-7], 21:58; 24:572
   in brazing filler metals, 22:494
   refractory coatings for, 21:105
   role in steelmaking, 22:779
   thin films of, 23:1060
   in tool steels, 24:398
   vapor pressure of, 23:1046
Tungsten and tungsten alloys, 24:**572**
Tungsten borides, 24:597
   in refractory coatings, 21:91
Tungsten carbide [12070-12-1], 21:847;
      24:404, 597
   electroplated thin films, 23:1072
Tungsten compounds, 24:**588**
Tungsten dibromide [13470-10-5], 24:591
Tungsten dichloride [13470-12-7], 24:590
Tungsten diiodide [13470-17-2], 24:591
Tungsten dioxide [12036-22-5], 24:592
Tungsten disilicide [12039-88-2], 24:598
Tungsten disulfide [12138-09-9], 24:596
Tungsten hexabromide [13701-86-5],
      24:591
Tungsten hexacarbonyl [14040-11-0],
      24:589
Tungsten hexachloride [13283-01-7],
      24:590
Tungsten hexafluoride [7783-82-6],
      24:589
   for chemical vapor deposition, 23:1060
Tungsten nitride [12058-38-7], 24:597
Tungsten oxychloride, 23:663
Tungsten oxydibromide [13520-75-7],
      24:591
Tungsten oxydichloride [13520-76-8],
      24:590
Tungsten oxydifluoride [14118-73-1],
      24:590
Tungsten oxydiiodide [14447-89-3], 24:591
Tungsten oxytetrabromide [13520-77-9],
      24:591
Tungsten oxytetrachloride [13520-78-0],
      24:590
Tungsten oxytetrafluoride [13520-79-1],
      24:590

Tungsten oxytrichloride [14249-98-0], 24:590

Tungsten pentabromide [13470-11-6], 24:591

Tungsten pentachloride [13470-13-8], 24:590

Tungsten pentafluoride [19357-83-6], 24:590

Tungsten–rhenium alloys, 21:339

Tungsten selenide
  as refractory coating, 21:113

Tungsten tetrabromide [12045-94-2], 24:591

Tungsten tetrachloride [13470-14-9], 24:590

Tungsten tetrafluoride [13766-47-7], 24:590

Tungsten tetraiodide [14055-84-6], 24:591

Tungsten tribromide [15163-24-3], 24:591

Tungsten triiodide [15513-69-6], 24:591

Tungsten trioxide [1314-35-8], 24:592

Tungsten trisulfide [12125-19-8], 24:596

Tungstic acid [7783-03-1], 24:593

Turbidimetry
  use in particle size measurement, 22:269

Turbidity and nephelometry, 24:602

Turbine components
  brazing filler metals for, 22:490

Turbine mills, 22:291

Turbines
  steam, 22:746
  for wind energy, 22:467

Turkey red oil, 24:602

Turnips
  sulfoxides in, 23:217

Turpentine [8006-64-2], 24:602
  sampling standards for, 21:628
  as solvent, 22:548

Turpentine oil
  specific gravity, 23:625

Tween
  sorbitan surfactant, 23:516

Twin-fluid atomizer, 22:672

Twitchell's reagents, 23:211

Two-frequency differential absorption lidar, 22:648

Two-shot method
  silicate grouting, 22:453

Tychite, 22:404

Tylosin [1401-69-0], 24:827, 829

Tyndalization, 22:847

Tyndall scattering, 22:628

Type 4340 alloy steel
  transformation diagram, 22:797

Type metal, 24:602

Type V gel–silica, 22:519

Typhoid
  vaccine against, 24:742

Tyrocidine, 24:602

Tyrosine
  in silk, 22:156

Tyrothricin, 24:602

Tyuyamunite [12196-95-1], 24:643

Tyvek–Mylar, 23:553

Tyvek–paper, 23:553

TYZOR AA [17927-72-9], 24:291

TYZOR AA95 [9728-09-9], 24:292

TYZOR DC [27858-32-8], 24:293

TYZOR ET [3087-36-3], 24:276

TYZOR IBAY [83877-91-2], 24:293

TYZOR ISTT [61417-49-0], 24:330

TYZOR KTM [83897-99-8], 24:276

TYZOR LA [65104-06-5], 24:291

TYZOR NPT [3087-37-4], 24:278

TYZOR TBT [5593-70-4], 24:275

TYZOR TE [36673-16-2], 24:295

TYZOR TOT [1070-10-6], 24:276

TYZOR TPT [546-68-9], 24:275

U

Ubbelohde viscometer, 21:376

Ucarsol LE, 23:440

Udex process, 24:361

UIC. See Underground-injection controls.

Ultra-accelerators
  in latex products, 21:584

Ultracentrifuges, 21:852
  pattern for soybeans, 22:594

Ultrafiltration, 24:603, 743
  pretreatment for membrane feed, 21:316

Ultraforming process, 24:359

Ultramarines, 21:980; 24:626

Ultrapure water
  from reverse osmosis, 21:326

Ultrasonic atomizer, 22:673

Ultrasonic devices
  as repellents, 21:260

Ultrasonic generators
  synthetic quartz crystals for, 21:1082

Ultrasonics
    use in particle size measurements,
        22:265
Ultrasonic spectroscopy
    for particle size measurement, 22:272
Ultratrace analyses, 24:491
Ultraviolet absorbers, 24:626
Ulysses, 23:1036
Umber, 24:626
Unads, 24:626
1-Undecanethiol [5332-52-5], 24:21
Undecapeptide substance P
    immunometric assay of, 24:508
Underground-injection controls (UIC),
        21:167
Underground storage tank (UST), 21:167;
        24:626
Uniaxial viscosity, 21:365
UNICARB, 23:465
United States Pharmacopeia, 24:516
    classification of sutures, 23:541
Units and conversion factors, 24:**627**
Unsaturated polyester resins (UPR),
        22:984
Unsaturated polyesters, 24:638
UNS designation
    for steel, 22:819
UPR. See Unsaturated polyester resins.
Urania
    in refractories, 21:78
Uraninite [1317-99-3], 24:642, 643, 648
    magnetic intensity, 21:876
Uranium [7440-61-1], 24:638
    analysis by x-ray absorption, 22:655
    enrichment, 24:656
    exposure to, 24:684
    extraction from spent fuel rods, 23:468
    in geological studies, 24:639
    isotopes, 24:639
    metal, properties of, 24:651
    in nuclear reactors, 24:660
    occurrence in nature, 24:641
    organometallic complexes, 24:679
    prdn, 24:646
    radioactive decay products, 24:639
    recovery from ores, 24:646
    resources, 24:644
    sampling standards for, 21:628
    separation by reverse osmosis, 21:323
    sulfur in ore leaching, 23:251
    thorium in, 24:70
    use of sulfuric acid, 23:397
    vitreous silica impurity, 21:1036

Uranium and uranium compounds, 24:**638**
Uranium carbide [12070-09-6], 24:665
Uranium carbonates, 24:668
Uranium dicarbide [12071-33-9], 24:665
Uranium dioxide [1344-57-6], 24:638,
        651, 662
Uranium hexachloride [13763-23-0],
        24:678
Uranium hexafluoride [7783-81-5], 24:651,
        660, 677
    centrifugal separation of, 21:871
    sampling standards for, 21:628
Uranium mononitride, 24:664
Uranium nitride [25658-43-9], 24:664
Uranium oxide [1344-58-7]
    as refractory material, 21:57
Uranium oxides, 24:661
Uranium pentabromide [13775-16-1],
        24:679
Uranium pentachloride [13470-21-8],
        24:678
Uranium pentafluoride [13775-07-0],
        24:677
Uranium phosphates, 24:670
Uranium tetrabromide [13470-20-7],
        24:679
Uranium tetrachloride [10026-10-5],
        24:638, 678
Uranium tetrafluoride [10049-14-6],
        24:651
Uranium tetraiodide [13470-22-9], 24:679
Uranium tribromide [13470-19-4], 24:678
Uranium triiodide [13775-18-3], 24:678
Uranium trinitride [12033-85-1], 24:664
Uranium trioxide [1344-58-7], 24:651
    prprn of, 24:663
Uranophane [12195-76-5], 24:643
Uranyl acetate
    in sodium ore, 22:342
Uranyl nitrate [10102-06-4], 24:666
Uranyl nitrate hexahydrate [13520-83-7],
        24:651
Urea-adduction method, 23:138
Urea-formaldehyde
    in textile finishing, 23:896
Urea-formaldehyde resins, 24:694
    in grouting systems, 22:457
Urease
    in soybeans, 22:596
Urethane grouts, 22:456
Urethane polymers, 24:**695**
Urethane sealants, 21:657

Urinary calculi
  trace analysis of, *24*:518
Urine
  analysis of heavy metals in, *22*:646
  trace analysis of, *24*:508
Urons
  in textile finishing, *23*:899
Used oil, *21*:1
UST. See *Underground storage tank.*
Uv stabilizers
  selenium as, *21*:713
  in styrene plastics, *22*:1058

V

Vaccines, *24*:622
  separations for prdn of, *21*:855
  as veterinary drugs, *24*:837
Vaccine technology, *24*:**727**
Vaccinology, *24*:743
Vacuum applications, *24*:751
Vacuum bag molding, *21*:196
Vacuum drying
  of soap, *22*:314
Vacuum dynamics, *24*:753
Vacuum evaporation, *23*:1044
Vacuum melting
  for steel melting, *22*:771
Vacuum metallizing boats
  titanium diboride in, *24*:228
Vacuum systems
  diffusion pump, *24*:760
  gas transport, *24*:764
  leaks, *24*:762
  microstructure on surfaces, *24*:759
  turbulent gas flow, *24*:760
  wall materials, *24*:773
Vacuum technology, *24*:**750**
  for film formations, *23*:1041
  vitreous silica pipes for, *21*:1065
Valine
  in silk, *22*:156
  in soybeans and other oilseeds, *22*:596
Valves
  steel for, *22*:822
Vanadic acid, meta [*13470-24-1*], *24*:800
Vanadinite [*1307-08-0*], *24*:783
Vanadium [*7440-62-2*], *24*:782, 797
  by acid leaching, *24*:806
  catalysts for sulfuric acid manufacture, *23*:390

interstitial compounds, *24*:801
recovery, *24*:805
role in steelmaking, *22*:779
in steels, *22*:811
sulfur in ore leaching, *23*:251
thorium oxides, *24*:76
in tool steels, *24*:399
use of sulfuric acid, *23*:397
Vanadium(III) acetylacetonate [*13476-99-8*], *24*:800
Vanadium and vanadium alloys, *24*:**782**
Vanadium carbide [*12070-10-9*], *24*:800
Vanadium carnotite [*60182-49-2*], *24*:643
Vanadium compounds, *24*:**797**
  color, *24*:801
  magnetism, *24*:801
  by salt roasting, *24*:804
Vanadium(IV) disilicide [*12039-87-1*], *24*:800
Vanadium halides, *24*:803
Vanadium nitride [*24646-85-3*], *24*:800
Vanadium(II) oxide [*12035-98-2*], *24*:800
Vanadium(III) oxide [*1314-34-7*], *24*:800
Vanadium(IV) oxide [*12036-21-4*], *24*:800
Vanadium(V) oxide [*1314-62-1*], *24*:800
Vanadium oxyhalides, *24*:803
Vanadium(V) oxytrichloride [*7727-18-6*], *24*:800, 802
Vanadium–silicon alloy, *21*:1120; *24*:786
Vanadium steels, *24*:794
Vanadium(III) sulfate [*13701-70-7*], *24*:803
Vanadium(IV) tetrachloride [*7632-51-1*], *24*:800
Vanadium tetrachloride [*7632-51-1*], *24*:802
Vanadium–tin yellow, *24*:127
Vanadium tribromide [*13410-26-3*], *24*:803
Vanadium trichloride [*7718-98-1*], *24*:800, 803
Vanes
  refractory coatings for, *21*:106
Vanilla
  extraction using SCFs, *23*:466
Vanilla beans, *24*:812
Vanillin [*121-33-5*], *23*:205; *24*:**812**
  from lignosulfonate, *23*:169
Vanillin sugar, *24*:821
Vaniltek, *24*:823
Vanthoffite, *22*:404
van't Hoff's equation, *21*:311
Vapor crystal growth, *22*:624

Vapor degreasing, 22:569
Vapor–liquid equilibria, 24:825
Vapor–phase epitaxy, 23:1060
Vapor pressure
    of steam, 22:721
Varamide
    diethanolamine surfactant, 23:520
Varine
    surfactant, 23:527
Varion
    surfactant, 23:531
Variquat
    surfactant, 23:529
Varisoft
    surfactant, 23:529
Varistors
    titanium dioxide in, 24:238
Varnishes, 24:825
    purification by centrifuge, 21:857
    sampling standards for, 21:628
Varox 365
    amine oxide surfactant, 23:525
Varox 185E
    amine oxide surfactant, 23:525
Vat Blue 7 [6505-58-4], 23:354
Vat Blue 42 [1327-81-7], 23:342, 353
Vat Blue 43 [1327-79-4], 23:342
Vat dye [1328-11-6], 23:354
Vat Green 7 [1328-12-7], 23:354
Vat Orange 21 [1328-39-8], 23:354
Vat Yellow 21 [1328-40-1], 23:354
VDC. See Vinylidene chloride.
VDC copolymer latices, 24:916
Vebe process, 23:739
Vegetable fibers, 24:825
Vegetable oils
    fractionation using SCFs, 23:466
    purification by centrifuge, 21:857
    separation of acids from, 21:926
    sulfurization of, 23:429
Vegetable products
    sorbates in, 22:583
Vegetable waxes
    as release agents, 21:210
Velvetex, 24:825
    surfactant, 23:531
Vendex, 24:139
Veramine [21059-48-3], 22:863
Verbascose
    in soybeans and other oilseeds, 22:598
Verdefilm, 21:715
Vermiculite, 24:825

Vermilion, 24:825
Vernolate
    thiols in, 24:30
Versatate VV9, 24:946
Versatate VV10, 24:946
Vertifoam process, 24:711
Veterinary drugs, 24:**826**
    DMSO as, 23:227
Veterinary products
    regulatory agencies, 21:177
Vetiver, 24:838
Vezin sampler, 21:645
Vibrational circular dichroism, 22:651
Vibration control
    shape-memory alloys for, 21:972
Vibrations
    of silanol groups, 22:517
Vibrations, molecular
    spectroscopy of, 22:636
Vibratory atomizer, 22:673
Vibrio parahaemolyticus
    inhibited by sorbates, 22:580
Vickers hardness
    of vitreous silica, 21:1054
Vicker's hardness
    of cermets, 24:415
Vicryl, 23:543
Vicryl Rapide, 23:548
Victoria Blue B [2580-56-5], 24:560
Vidal Black
    sulfur dye, 23:342
Video cameras
    sensors for, 21:820
    solid tantalum capacitors in, 23:670
Video compact disks, 22:176
Vidicons
    use of selenium, 21:708
Vifilcon-A, 24:1079
Viking, 23:1036
Vinblastine [865-21-4], 24:836
Vincristine [57-22-7], 24:836
Vinegar, 24:**838**
    dextrose for, 23:592
    sorbitol in, 23:104
    specific gravity, 23:625
Vinyl acetal polymers, 24:924
N-Vinylacetamide [5202-78-8], 24:1071
N-Vinylacetanilide [4091-14-9], 24:1071
Vinyl acetate [108-05-4], 24:943
    copolymerization with VDC, 24:888
    VP copolymerization, 24:1090
Vinylacetylene [689-97-4], 24:856

Vinylbenzene, *24*:851

Vinylbenzyl cationic silane [*34937-00-3*]
silane coupling agent, *22*:150

N-Vinylcaprolactam [*2235-00-9*], *24*:1071
VP copolymerization, *24*:1090

Vinyl chloride [*75-01-4*], *24*:**851**, 1043
copolymerization with VDC, *24*:888
regulatory level, *21*:160

Vinylcopper [*37616-22-1*], *24*:854

Vinylec, *24*:936

Vinyl ester
silane coupling agent for, *22*:152

Vinyl ether, *24*:882

Vinyl fibers, *24*:882

Vinyl fluoride [*75-02-5*], *24*:854

Vinylidene [*2143-69-3*], *24*:856

Vinylidene chloride (VDC) [*75-35-4*],
24:882, 884

Vinylidene chloride monomer and
polymers, *24*:**882**

Vinylidene polymers, *24*:923

Vinyl ketal polymers, *24*:924

Vinyllithium [*917-57-7*], *24*:854

Vinylmagnesium chloride [*3536-96-7*],
24:853

N-Vinyl-5-methyl-2-oxazolidinone
[*3395-98-0*], *24*:1071

(*S*)-5-Vinyl-2-oxazolidinethione [*500-12-9*],
24:100

N-Vinyl-2-oxazolidinone [*4271-26-5*],
24:1071

p-Vinylphenol [*2628-17-3*], *24*:856

N-Vinylphthalimide [*3485-84-5*], *24*:1071

N-Vinyl-2-piperidinone [*4370-23-4*],
24:1071

Vinyl acetal polymers, *24*:**924**

Vinyl acetate polymers, *24*:**943**

Vinyl alcohol polymers, *24*:**980**

N-Vinylamide polymers, *24*:**1070**

Vinyl chloride polymers, *24*:**1017**

Vinyl ether monomers and polymers,
24:**1053**

Vinyl polymers, *24*:**924**

N-Vinyl-2-pyrrolidinone (VP) [*88-12-0*],
24:1071

Vinylsilane [*7291-09-1*], *22*:42

Vinyl stearate
thin films of, *23*:1084

Vinyltoluene [*25013-15-4*], *22*:985; *24*:386,
1099

Vinyltrichlorogermane [*4109-83-5*], *24*:854

Vinyltrichlorosilane [*75-94-5*], *24*:854

Vinyltriethoxysilane
critical surface tension of, *22*:148

Vinyltrimethoxysilane [*2768-02-7*]
silane coupling agent, *22*:150

Vinyltris(isopropenoxy)silane [*15332-99-
7*]
as silicone cross-linker, *21*:655

Violet BNP [*80539-34-0*], *24*:564

Viral infections, *24*:1099

Viral influenza
vaccines against, *24*:733

Virginiamycin [*11006-76-1*], *24*:829

Viruses
centrifugal separation of, *21*:852

Viscoelastic materials, *21*:347

Viscometers, *21*:347

Visco-Mix, *21*:396

Viscosity, *21*:347

Viscosity number, *21*:356

Viscotester, *21*:396

Viscotron, *21*:394

Viscous flow, *21*:347

Vista
alkylbenzensulfonate surfactant, *23*:496

Vistalon
ethylene–propylene polymer, *21*:486

Vitamin A [*68-26-8*], *24*:829

Vitamin B$_1$, *24*:500

Vitamin C
from sorbitol, *23*:111

Vitamin D, *22*:852

Vitamin D$_2$ [*50-14-6*], *22*:856

Vitamin D$_3$ [*67-97-0*], *22*:856

Vitamin E
SCF extraction of, *23*:467

Vitamins
dextrose in prdn of, *23*:592

Vitavax
thiols in, *24*:30

Viton, *24*:32

Vitreosil, *21*:1058

Vitreosil-ir, *21*:1034

Vitreosil-O55, *21*:1034

Vitreous refractory fibers, *21*:121

Vitreous silica [*60676-86-0*], *21*:1032

Vitrification
effect of supercritical fluid on, *23*:460

VM&P naphtha
as solvent, *22*:538

VOC. See *Volatile organic compounds.*

VOC emissions reduction
and solvents, *22*:531

Voigt profile, *22*:632
Volatile organic carbon emissions
  reduction of using SCFs, *23*:468
Volatile organic compounds (VOC), *21*:167
Volpo
  ethoxylated alcohol surfactant, *23*:509
Voyager
  interferometric spectrometer spectra,
    22:642
  thermoelectric generators for, *23*:1035
VP. See *N-Vinyl-2-pyrrolidinone.*
Vrbaite [*12006-31-4*], *23*:952
Vulcanization, *22*:51
  accelerators of, *21*:461
  models for, *23*:429
  of natural rubber, *21*:572
Vulcanized silicone rubber, *22*:110
Vultacs, *23*:288

W

Wackenroeder's liquid, *23*:278
Wagner equation, *23*:990
Wallace plasticity
  of rubber, *21*:572
Wall geometries, *24*:767
Wallpaper pastes
  starch in, *22*:712
Washcoats, *22*:696
Washing
  in paper recycling, *21*:15
Wash primers
  PVB in, *24*:935
Wasserglass, *22*:1
Waste activated sludge
  centrifugal separation of, *21*:850
Waste minimization
  effect of SCFs on, *23*:467
Waste oil, *21*:1
Waste reduction
  use of steam, *22*:760
Waste rubber products, *21*:35
Wastes
  tanks for, *23*:644
Waste streams
  hydrothermal oxidation of, *23*:470
Waste-to-energy (WTE), *21*:227
Waste treatment
  silicates for, *22*:22
  sludge centrifugation, *21*:852

Wastewater
  sampling, *21*:642
Wastewater treatment. (See also *Water treatment.*)
  sodium tetrasulfide in, *22*:418
  using SCFs, *23*:468
Watches
  solar energy for, *22*:475
  synthetic quartz crystals for, *21*:1082
Water [*7732-18-5*], *21*:930
  additives, *22*:18
  analyses in steam prdn, *22*:745
  diffusion in vitreous silica, *21*:1047
  refractories for, *21*:79
  role in sulfur prdn, *23*:240
  role in temperature measurement, *23*:812
  sampling standards for, *21*:628
  as SCF, *23*:470
  specific gravity, *23*:625
  steel quenching in, *22*:801
  supercritical fluid, *23*:454
  tanks for, *23*:644
  ultrafiltration, *24*:622
Water, pure
  boiling point of, *23*:627
Water, sea. See *Seawater.*
Water-based coatings, *22*:568
Water-based laminating inks
  titanate adhesion promoters in, *24*:329
Water-borne polyurethane coatings, *24*:716
Water clarification
  in paper mills, *21*:19
Water permeability coefficient
  role in reverse osmosis, *21*:309
Water pollution control, *21*:154
Waterproofing, *23*:907
  silylating agents for, *22*:47, 148
Water purification
  supercritical fluid extraction, *23*:465
Water repellency
  from titanates, *24*:290
Water repellents
  silicon esters in, *22*:79
  silylating agents as, *22*:148
  thiols for, *24*:30
  titanates as, *24*:324
Water softeners
  sodium chloride in, *22*:373
Water-soluble fluxes, *22*:496
Water-soluble silanols, *22*:127

Water-treating compounds
  tanks for, *23*:644
  use of sulfur for, *23*:252
Water treatment
  role in steam prdn, *22*:738
  role of sedimentation in, *21*:672
  silicates for, *22*:21
  silver for, *22*:167
  sodium bromide for, *22*:379
  sodium sulfite for, *23*:314
  swimming pools, *22*:379
Wave energy, *21*:234
Waveguide fibers
  by sol–gel technology, *22*:500
Waveguides
  compound semiconductors, *21*:788
  tantalates in, *23*:677
  thorium compounds as, *24*:77
  vitreous silica fibers, *21*:1040
Waxes
  as paper contaminants, *21*:11
  as release agents, *21*:207
  as suture coating, *23*:550
  triarylmethane dyes for, *24*:565
Waxy corn
  starch from, *22*:699
Weatherproofing
  sealant use, *21*:665
Weber numbers
  in sprays, *22*:684
Weight
  measurement of, *24*:633
Weight reduction
  stimulants for, *22*:937
Weissenberg effect, *21*:371
Welding
  steel equipment for, *22*:823
  use of tellurium in, *23*:802
Wellman-Lord process, *23*:448
Wells
  sulfamic acid cleaner for, *23*:129
Wells-Brookfield viscometer, *21*:395
Wet ball mill, *22*:294
Wet etching
  compound semiconductor processing,
    21:798
Wet grinding, *22*:294
Wet system compression molding, *21*:200
Wetting agents
  in adhesives, *24*:971
  lignosulfates as, *23*:205
Whale oil
  specific gravity, *23*:625

Wheat
  dextrose from, *23*:585
  starch from, *22*:699
Wheat stem rust
  sugar alcohols in, *23*:96
Whey
  desalting of, *21*:327
Whiskers
  in tool materials, *24*:435
White smoke
  titanium tetrachloride as, *24*:258
Whole-tree-energy plants, *21*:227
Whooping cough
  vaccines against, *24*:728
Wilkinson's catalyst, *22*:55
Williams-Landel-Ferry equation, *21*:356
Williamson equation, *21*:350
Wind
  as solar energy, *22*:466
Windows
  PVC use, *24*:1040
Windows, double-paned
  silver coatings for, *22*:176
Windshields
  PVB interlayer, *24*:932
Wind turbines, *21*:225
Wine, *21*:1020; *24*:842
  sorbitol in, *23*:96, 104
  use of sorbates, *22*:582
  use of sulfur dioxide in mfg, *23*:310
Winstrol-V, *24*:834
Wiping stains, *22*:696
Wire
  PVC use, *24*:1040
  tantalum powders for, *23*:669
  tire cords, *21*:1116
Wire enamel
  PVF resins, *24*:938
Witcamide
  diethanolamine surfactant, *23*:520
Witcamide 70
  surfactant, *23*:520
Witcamide CPA
  surfactant, *23*:520
Witcamine
  surfactant, *23*:527
Witcolate
  alkylphenol surfactant, *23*:503
Witcolate AM
  surfactant, *23*:501
Witcolate D-510
  surfactant, *23*:501

Witcolate TLS500
  surfactant, 23:501
Witcolate WAC-LA
  surfactant, 23:501
Witconate
  alkylbenzensulfonate surfactant, 23:496
Witconol
  glycerol ester surfactant, 23:512
  polyoxyethylene surfactant, 23:514
Witconol CAD
  fatty acid surfactant, 23:518
Withanolide D [30655-48-2], 22:867
Withanolide E [38254-15-8], 22:867
Withanolides, 22:852
Withering
  unit operation of tea processing, 23:757
Wodginite [12178-62-0]
  tantalum in, 23:660
Wolframite
  magnetic intensity, 21:876
Wollastonite [14567-51-2], 22:3
Wood, 23:96
  digestion, 22:417
  sodium bromide preservative, 22:379
  thioglycolic acids for, 24:15
Wood alcohol
  boiling point of, 23:627
Wood chips
  for silicon prdn, 21:1105
Wood fillers
  in staining process, 22:696
Wood finishing, 22:692
Wood fuel, 21:227
Wood pulp
  sulfur in prdn of, 23:259
Wood pulping
  reverse osmosis for, 21:327
Wood rosin process, 21:292
Wood shakes
  on roofs, 21:453
Wood shingles
  on roofs, 21:453
Wood stains, 22:692
Wool
  sampling standards for, 21:628
  sulfonic acid derived dye for, 23:207
  titanated inks for, 24:329
  treatment by thioglycolic acid, 24:15
  use of ultrafiltration, 24:622
  vitreous silica, 21:1069
Wool Fast Blue FBL [6661-40-1], 24:556
Wool grease
  centrifugal separation of, 21:860

Wootz, 22:766
Worker protection, 21:150
Work index
  particle size reduction, 22:281
Worthite, 23:307
Woven fabrics
  in reinforced plastics, 21:195
Wrist watches
  silver cmpds for batteries, 22:190
WTE. See Waste-to-energy.
Wulfenite [14913-82-7], 24:643
Wurtz-Fittig coupling
  of organosilanes, 22:63
Wurtz reaction, 24:134

X

Xanthates
  vulcanization accelerator, 21:463
Xenon [7440-63-3]
  diffusion in vitreous silica, 21:1048
  supercritical fluid, 23:454
Xenon arc lamp
  for spectrofluorometers, 22:652
Xenotime
  magnetic intensity, 21:876
Xerogels, 21:996, 1020; 22:25
  alumina, 22:521
  by sol–gel technology, 22:500
Xeroradiography
  use of selenium, 21:709
XPS. See Dibutylxanthogen polysulfide.
X-ray diffraction
  silk structure defn, 22:157
X-ray fluorescence, 22:629, 655
X-rays
  for particle detection, 22:268
X-ray spectroscopy, 22:654
Xylan, 23:96
Xylazine hydrochloride [23076-35-9],
  24:834
Xylene [1330-20-7]
  separation by reverse osmosis, 21:323
  as solvent, 22:540
  sulfonation of, 23:159
m-Xylene [108-38-3]
  as HAP compound, 22:532
  as solvent, 22:540
o-Xylene [95-47-6]
  as HAP compound, 22:532
  as solvent, 22:540

p-Xylene [*106-42-3*]
   centrifugal separation of, *21*:867
   as HAP compound, *22*:532
   as solvent, *22*:540
m-Xylene-*m*-sulfonic acid, *23*:203
Xylitol [*87-99-0*], *23*:95
   as sweeteners, *23*:556
m-Xylylene diisocyanate [*3634-83-1*],
   24:707

Y

Yarns, *23*:886
   sampling standards for, *21*:628
   use of ultrafiltration, *24*:622
Yeasts, *23*:98
   centrifugal separation of, *21*:860
   inhibited by sorbates, *22*:579
Yellow prussiate of soda
   in dendritic salt prdn, *22*:363
Yield stresses, *21*:354
Young's modulus
   of vitreous silica, *21*:1042, 1053
Yttria, *21*:84
Yttrium aluminum garnet
   in tool materials, *24*:432
Yttrium oxide
   in tool materials, *24*:431
Yttrotantalite [*12199-77-8*]
   tantalum in, *23*:660

Z

Zearalenol, *22*:599
Zearalenone, *22*:599; *24*:500
Zeeman spectra, *22*:636
Zeolites, *21*:980
   in benzene alkylation, *22*:961
   inorganic silicates, *22*:25
   silicates for, *22*:21
   titanium silicates in, *24*:264
Zeranol [*26538-44-3*], *24*:834
Zero-insertion-force connector
   shape-memory alloy as, *21*:969
Zero-shear viscosity, *21*:359
Zest, *22*:323
Zeta potential, *22*:522
Ziegler-Natta catalysts, *24*:82
   effect of silylating agents on, *22*:147
   titanium trichloride in, *24*:257

Ziegler-Natta polymerization, *24*:310
Zinc
   adhesion in tire cord, *24*:171
   in brazing filler metals, *22*:492
   plating with methanesulfonic acid,
     23:326
   in scrap for steel, *22*:779
   separation by reverse osmosis, *21*:322
   in solders, *22*:485
   strontium carbonate in prdn of, *22*:953
   sulfuric acid as by-product, *23*:381
   vapor pressure of, *23*:1046
   vitreous silica impurity, *21*:1036
Zinc alloys
   as shape-memory alloys, *21*:964
Zincblende semiconductors, *21*:763
Zinc carbonate [*12122-17-7*]
   as rubber chemical, *21*:473
Zinc diamyldithiocarbamate [*15337-18-5*],
   21:466
Zinc dibenzyldithiocarbamate [*14726-36-
   4*], *21*:466
Zinc di-*n*-butyldithiocarbamate [*136-23-2*],
   21:466
Zinc *O,O*-di-*n*-butylphosphorodithioate
   [*6990-43-8*], *21*:468
Zinc diethyldithiocarbamate [*14324-55-1*],
   21:466
Zinc diisobutyldithiocarbamate [*36190-
   62-2*], *21*:466
Zinc diisononyldithiocarbamate, *21*:466
Zinc diisopropylxanthate [*1000-90-4*],
   21:468
Zinc dimethyldithiocarbamate [*137-30-4*],
   21:466
Zinc dithionite [*7779-86-4*], *23*:319
Zinc 2-ethylhexoate [*136-53-8*]
   as rubber chemical, *21*:473
Zinc formaldehyde sulfoxylate [*24887-
   06-7*], *23*:319
Zinc laurate [*68242-42-0*]
   as rubber chemical, *21*:473
Zinc 2-mercaptobenzothiazole [*155-04-4*],
   21:465
Zinc oxide [*1314-13-2*]
   as rubber chemical, *21*:473
   sorbent for hydrogen sulfide, *23*:434
Zinc pentamethyldithiocarbamate
   [*13878-54-1*], *21*:466
Zinc process
   for prdn of sodium dithionite, *23*:317
Zinc salicylate, *21*:611

Zinc selenide
  substrate for self-assembled monolayer,
    23:1089
Zinc selenide [1315-09-9], 21:767
Zinc soaps, 22:324
Zinc–sodium alloys, 22:347
Zinc stearate [557-05-1]
  as release agent, 21:210
  as rubber chemical, 21:473
Zinc sulfide [1314-98-3], 21:767; 23:280,
    434
Zinc telluride [1315-11-3], 21:767
Zinc titanate
  for hydrogen sulfide removal, 23:435
Zipper effect, 24:14
Zircalloy-2, 24:121
Zircon [14940-68-2], 21:980; 22:3
  as refractory material, 21:50, 57
Zirconia, 21:55
  membrane for ultrafiltration, 24:606
  in sol–gel polymer hybrids, 22:526
  stabilized as coating, 21:113
  in tool materials, 24:430
Zirconia fibers, 21:124
Zirconia refractory fiber, 21:124
Zirconium, 23:394
  in brazing filler metals, 22:494

  in manganese–silicon alloys, 21:1118
  role in steelmaking, 22:779
  in shape-memory alloys, 21:970
  tin alloys of, 24:120
  vitreous silica impurity, 21:1036
Zirconium dioxide
  thin films of, 23:1058
Zirconium nitride
  in refractory coatings, 21:91
  thin films of, 23:1058
Zirconium oxide [1314-23-4], 21:119
  as refractory material, 21:57
Zirconium silicate, 21:54
Zirconium–silicon alloy, 21:1120
Zirconium tetrachloride
  tetrachlorosilane as by-product, 22:64
Ziziphin [73667-51-3], 23:577
ZK-98299 [096346-61-1], 22:903
Zoloft, 22:945
Zone refining
  of tellurium, 23:784
ZSM-5
  in benzene alkylation, 22:961
Zygosaccharomyces globiformis
  inhibited by sorbates, 22:580
Zygosaccharomyces halomembranis
  inhibited by sorbates, 22:580